HEPMAD24

HEPMAD24

Proceedings of the
14th High-Energy Physics International Conference

Antananarivo, Madagascar, 21 – 26 October 2024

edited by

Stephan Narison

Université de Montpellier II, France

World Scientific

NEW JERSEY • LONDON • SINGAPORE • BEIJING • SHANGHAI • HONG KONG • TAIPEI • CHENNAI • TOKYO

Published by

World Scientific Publishing Co. Pte. Ltd.
5 Toh Tuck Link, Singapore 596224
USA office: 27 Warren Street, Suite 401-402, Hackensack, NJ 07601
UK office: 57 Shelton Street, Covent Garden, London WC2H 9HE

British Library Cataloguing-in-Publication Data
A catalogue record for this book is available from the British Library.

HEPMAD24
Proceedings of the 14th High-Energy Physics International Conference

ISBN 978-981-98-2640-7 (hardcover)
ISBN 978-981-98-2641-4 (ebook for institutions)
ISBN 978-981-98-2642-1 (ebook for individuals)

For any available supplementary material, please visit
https://www.worldscientific.com/worldscibooks/10.1142/14679#t=suppl

Desk Editor: Rhaimie Wahap

Typeset by Stallion Press
Email: enquiries@stallionpress.com

COMMITTEE MEMBERS AND LOCAL ORGANIZATION

Chairman of the Conference
S Narison
(Laboratoire Univers et Particules, CNRS – UM – Montpellier — France)

International Committee
A Di Giacomo (Pisa)
M Knecht (Marseille)
A Pich (Valencia)
G Rossi (Roma)
V I Zakharov (Munich)

Local Organization
Hepmad Research Institute (iHEPMAD)
Association Gasy Miara-Mandroso (AGMM)

Conference Contact Information
Email: hepmad24@sciencesconf.org
Alternative Contact: snarison@yahoo.fr

PREFACE

HEPMAD 24 is the 14th International Conference on High-Energy Physics (HEP) held in Antananarivo - Madagascar initiated in 2001 with the collaboration of the Centre Européen de la Recherche Nuclaière (CERN) in Geneva, the University and INSTN of Antananarivo. It is an extension of the QCD-Montpellier Conference Series started in 1985. This 1st conference has involved about 30 participants from abroad and about 30 national physicists. It has been published by World Scientific Publishing Company (WSPC) based in Singapour (ISNB 981-238-086-8).

The contents of this 1st conference can be found from the link:

https://inspirehep.net/conferences/973738

Encouraged by the success of this 1st meeting, we have decided to continue the Series in 2004 after 3 years of thinkings for an initial 2 years frequency:

HEPMAD 04, 07, 09, 11, 13, 15

Since 2015, we organize the conference annually:

HEPMAD 16, 17, 18, 19.

The conference has been interrupted by the COVID19 pandemic and restarted in 2022.

The proceedings of theHEPMAD conferences from 2004 to 2022 have been published online on the eCONF platform of the SLAC-Stanford University (US). They can be consulted through the links below:

https://www.slac.stanford.edu/econf/C0409272/
https://www.slac.stanford.edu/econf/C0709107/
https://www.slac.stanford.edu/econf/C0908211/
https://www.slac.stanford.edu/econf/C110825/
https://www.slac.stanford.edu/econf/C130904/
https://www.slac.stanford.edu/econf/C150917/
https://www.slac.stanford.edu/econf/C161013/
https://www.slac.stanford.edu/econf/C170921/
https://www.slac.stanford.edu/econf/C180906/
https://www.slac.stanford.edu/econf/C191014/
https://www.slac.stanford.edu/econf/C221102/

The HEPMAD23 conference has not been published. We moved to the International Journal of Modern Physics of WSPC in 2024.

The main motivation for organizing this conference is to promote HEP in developing countries and especially in Madagascar where this branch of physics is unknown because it has been traditionnally dedicated to the developed countries.

Then, a such project becomes a great challenge as there are nor national experts or neither active practioners of this field while the large institutions such as Mundial Bank, European Committee,...and the Government only support applied science and technology while HEP is a fundamental research which studies the inner structure of matters (quarks and leptons) and the microscopic forces : nuclear mediated by gluons, electromagnetic (mediated by the photons and weak (mediated by the W and Z bosons).

A priori, a such field seems to be far from the actual needs of deveoping countries. However, this is not really true as any applied science and technologies need results and understanding from fundamental research. There are many indirect consequences of this high-energy physics research in the past and present technologies. Among these, one can quote:

- Electricity which we use everyday comes from the discovery of the electron (1st elementary particle) and from the understanding of its properties.
- TV cathodic screen is the smallest accelerator of particle (electron).
- Detector buildings stimulate high-technology in the industry and in other branches of physics (supra-conductors,...)
- Charpak type-detectors are used in the container-custom controls (le Havre),...
- The accelerator technology is used in medicine for e.g. the cancer treatment (Orsay, Geneva,...)
- The Internet Web pages have been found at CERN (Geneva) by Tim Berners-Lee for communicating rapidly the data between members of the experimental groups.
- High-Energy Physicists need powerful computers for treating data and for stimulating physical phenomena. In general, these experimental researches need high-technologies which are often at the level of prototypes and which will be useful in our everyday life in the near or long term future (cryogenics at -271^0 C,...).

In order to enlarge the impact of the conference to the country, the scientific program covered in the meeting is not only dedicated to HEP but is opened to different branches of physics researches practiced by national physicists (environment, climate, solar energy, chemical physics, ...). In this way, the national researchers take the opportunuity to publish their works at the international level. These publications are used by PhD students to validate their Thesis and by University Teachers for their career promotion such as the authorization to direct researchs (HDR).

Since the first conference, HEPMAD involves regularly about 5-15 oversea participants (especially HEP researchers) and about 20-30 national ones. The number has drastically decreased after the COVID 19 pandemics and the introduction of the on-line ZOOM conference.

An effort has been made to combine science with the discovery of the rare fauna and flora as well as some cultural aspects (musics, folks, arts) of Madagascar. However, the population spend hard time at present due to poverty, jobless, securities,... as a consequence of the bad governance by politicians in power.

I wish that most participants have spend a good week together in Antananarivo and discovered this Beautiful Country despite its catastrophic social situation.

I also wish that you have learned some new topics related to or beyond your current research fields.

Your are the ambassador of our conferences and we wish that you will transmit your experience from this meeting to your colleagues which may stimulate them to come and you to come back !

Cordially,

Stephan Narison (Chairman of the HEPMAD Conference Series)
E-mail of the conference: qcdconf@gmail.com
Url of the conference: https://www.lupm.umontpellier.fr/users/qcd/HEPMAD24

CONTENTS

---

[1] Manuscript not received.
[2] See the book: Particles and the Universe, ed. World Scientific Co., https://doi.org/10.1142/9454
[3] We notice that the Ministry of Energy and Mines, Ministry of Environment, Base toliara and INSTN have declined the invitation.

[4]These articles will not be included in this publication due to copyright restrictions.

[5]Manuscript not received.

Hepmad 24 Group Photos

Some Photos of the Organizers

Part 1

Opening Addresses and Public Talk

Stephan Narison
Chairman of the HEPMAD Conference,
Directeur de Recherche Emerite CNRS au Laboratoire Univers et Particules de Montpellier,
CNRS & Université de Montpellier, Place Eugéne Bataillon, 34095-Montpellier Cedex 05, France
E-mail of the conference: qcdconf@gmail.com Url of the conference: https://www.lupm.umontpellier.fr/users/qcd/HEPMAD24

Good morning to everybody and welcome to HEPMAD24,
On the behalf of the organizing committee, I would like to acknowledge all of you for your participation at the *14th HEPMAD International Conference on High-Energy Physics held in Antananarivo - Madagascar* and the personalities who came to honour this meeting especially Prof. Irene Rabenoro, President of the Malagasy Academy who will give the Official Opening of this HEPMAD24.

I also thank the spokespersons of different experiments who helped for preparing the program.

Last but not least, I thank the Institute iHEPMAD members of the University of Antananarivo and the "Association Gasy Miara-Mandroso" (AGMM) for their help in the logistics and for their cooperation in organizing this conference.

Motivated by promoting High-Energy Physics (HEP) in developing countries, especially in Madagascar, we have initiated the HEPMAD Series of Conferences in 2001 with the collaboration of the Centre Européen de la Recherche Nuclaière (CERN) in Geneva, the University and INSTN of Antananarivo.

After 3 years of thinkings and encouraged by the success of this 1st meeting, we have decided to continue the Series in 2004 every 2 years until 2015 and annually later on. Unfortunately, the pandemic OVID 19 has interrupted its organization which we have restarted in 2022.

This 1st conference has involved about 30 participants from abroad and about 30 national physicists. It has been published by World Scientific Publishing Company (WSPC) in Singapour (ISNB 981-238-086-8).

The other ones from 2004 to 2022 have been published on-line on the eCONF platform of the SLAC-Stanford University (US). They can be consulted through the links listed in the Preface of this proceedings.

The HEPMAD23 conference has not been published.

We moved to the International Journal of Modern Physics A of WSPC in 2024.
One should remind that HEP is a fundamental research, which studies the fundamental structure of matters and the origin of the universe such that its direct application to present technologies is non-trivial... . However, one can see some examples of its applications in the Preface.

For this reason, it becomes a challenge to introduce HEP in developing countries as the large institutions such as Mundial Bank, European Committee,...and the Government only support short term applied sciences and technologies.

Despite a such difficulty, the success of the HEPMAD Conference Series has encouraged us to continue the project which has been also stimulated by the interest of young universitaries to work in this field and by support of different colleagues.

Then, we founded in 2004 the iHEPMAD research institute at the University of Antananarivo with a branch for preparing Master and PhD in HEP.

Those who have chosen to study experiment have been hosted by the LHCb group at CERN while some went to the DUNE group at FERMILAB.

We have founded a theoretical group based in Antananarivo oriented to Quantum Chrodynamics (QCD), especially on the method of QCD spectral sum rules for studying the Hadron properties.

To better promote HEP, I give public talks at different institutions (Malagasy Academy, Universities, Alliance Française, Theological schools) and in different public and private High-Schools of Madagascar.

HEPMAD is an unique HEP Conference Series in Subsaharian and Indian Ocean countries. I wish that we will maintain its organization by your continuous participation.

Despite the difficult logistics (black out,...) and the small budget (conference fee of participants), we have tried to make the orgaization as good as we can. I wish that everybody present here will take benefit of the week spend together, both from the scientific and human point of views. I also wish that you will discover the beauty of this conutry despite its catastrophic social and political situation.

Good conference !..

Cordially,

Dr Stephan NARISON
QCD & HEP-MAD Conferences Chairman
Laboratoire Univers et Particules
CNRS / IN2P3 / UM2 Montpellier (FR)

Official Opening by the President of the Malagasy Academy[1]

[1] Manuscript not received.

Introduction to High-Energy Physics (HEP) and HEP in Madagascar[2]

S. Narison (LUPM Montpellier, FR and iHEPMAD, Univ. Antananarivo, MG)

[2]See the book: Particles and the Universe, ed. World Scientific Co., https://doi.org/10.1142/9454

Part 2

Debate on the Mine Project "Base Toliara"[3]

Chair : S. Narison

[3] We notice that the Ministry of Energy and Mines, Ministry of Environment, Base toliara and INSTN have declined the invitation.

The Base Toliara Project : Introduction to the Debate*

Stephan Narison

Laboratoire Univers et Particules de Montpellier (LUPM), CNRS-IN2P3, Case 070, Place Eugène Bataillon, 34095 - Montpellier, France
and
Institute of High-Energy Physics of Madagascar (iHEPMAD) University of Ankatso, Antananarivo 101, Madagascar

Abstract

This Special Session of the HEPMAD 24 Conference is a Debate on the project for extracting mineral sands which contain Ilmentite, Monazite and Rutile for producing rare earths initiated by the Australian Company "Base Toliara" and recently bought by the American Company Energy Fuels Inc. This exploitation has been suspended in 2019 due to its negative impact on bio-diversity and eco-systems (flora and fauna rare species,,,,) and on the population (health caused by radioactivity, lands,...). However, motivated by its supposedly positive economical impact, it has been re-opened by the Malagasy Government in November 2024 despite the dissentiment of various associations, political personalities and ordinary citizens.

Keywords: Nuclear Physics, Radioactivity, Environment, Society.

*Talk given at the 14th International Conference in High-Energy Physics - HEPMAD24, 21-26th october 2024, Antananarivo-MG.

Email address: `snarison@yahoo.fr` (Stephan Narison)

Preprint submitted to Review of Modern Physics A *February 19, 2025*

1. Introduction

The Base Toliara project is an exploitation of mineral sands which contain Ilmentite, Monazite and Rutile for producing rare earths. It is situated in the Region of the South West of Madagascar has an area of about 470 km^2 and located about 80 km from Tulear city. Its directly impacts on about 200 000 habitants distributed over 12 municipalities, on the rare MIKEA tribe and on environment (rare species of fauna and flora due to the destruction of about 400ha of forest). Initiated by the Australian Company "Base Toliara", the project has been suspended in 2019 due to its negative impact on environment and population health and lands. It has been recently bought by the American Company Energy Fuels Inc. Motivated by its supposedly positive economical impact, it has been re-opened by the Malagasy Government in November 2024 despite the dissentiment of various associations, artists, deputies, political parties and citizens.

To have unbaised debate in this conference, we have equally invited:

◇ Some representants of the Pro-project (Base Toliara, "Olobe (respectable personalities)" Platform, Ministry of Mines and Energy, Minisitry of Environment) and of the National Institute of Nuclear Sciences and Technologies (INSTN) who have favourably confirmed the non-dangerous rating of the radioactivity implied by the project.

◇ Some members of the Civil Societies and Associations against the Project.

◇ Some Personalities and Ordinary Citizens.

Unfortunately, none of the Pro-project and INSTN members have replied to the invitation.

2. Radioactivity impact on environment and population

◇ It is known that the lmentite, Monazite and Rutile contains heavy radioactive mineral (Uranium-238, Thorium-232 and Potassium-40) which become active once released on the surface in the contact with air. An analysis of a sample of this sand using apparatus of the INSTN of Antananarivo has shown[1] that these minerals contain a radioactivity rate of about 10600 Bequerel/kg which is far above the tolerated by the human body of 10 000 Bequerel/kg. An independant analysis by the INSTN members claimed (without any available scientific publications) that the previous analysis is wrong [2][1]. The government uses a such result for authorizing the re-opening of the Base Toliara project.

[1]We regret that the members of the INSTN have declined the inivitation to present their results at the conference.

◇ However, because of these two contradictory conclusions from Ref. [1] and the INSTN, it is mandatory to have a new independent unbiaised, transparent analysis with some open-access results of different samples without any political influence and pressure. This request is one of the scientific point raised in this Debate.

◇ One should note that in the South East part of Madagascar a similar exploitation is being done near Fort-Dauphin city by the Quit Madagascar Mineral (QMM), a branch of Rio Tito Iron and Titanium (RTIT) of Quebec-Canada, where the rate of radioactivity is much less (181-212 Becquerel/kg) [1]. However, on one hand, some civil societies, associations and journalists (Taniko, Craad-oi, Andrew Lees Trust,..) have observed after several years of the mine exploitation its negative impact on environment (several fishes are abnormally dead near the site, the number of fishes deceases as noted by some national fishers,..) and on the health of the population (new pathologies for children and old persons,...). Then, after a long term exploitation which is expected to be beyond 2054 for the Base Toliara, one would expect a tragic situation for the new generation of population and for the environment caused by radioactivity. One should note that the promise about the security announced in the contract is not often respected by Multinational Societies. This non-respect is often accompained by the connivance with the security controllers.

◇ On the other hand, the supposedly development of the Region of Fort-Dauphin brought by the mineral exploitation of the QMM society is quite marginal. It favours mainly the Malagasy employers in that society, the subcontracting companies and the personalities of the government who have signed the contract through some commissions.

3. Some other Negative Impacts on Environment and Population

Here, I shall only give below a list of some items where some of them will be developed in the talks of Z. Randriamaro [3] and M. Rakotondrainibe [4]. They are the:

◇ Destruction of 400 ha of forest which will destroy the rare fauna and flora of this region and will impact on the eco-systems and the change of climate which is drastic as the South West of Madagascar has an arid climate. The presence of this forest which is the lung of this region is mandatory for its survival.

◇ Disappearance of the MIKEA Tribe. Note that this tribe is among the oldest and rarest in the World similar tto the Aboriginal of Australia and needs to be preserved.

◇ Land grabbing from farmers despite the government's promise. The grabbing of the land is not done with the agreement and the wish of the owners but within some intimidations while the value of the land is often underestimated and often not payed. As a consequence, the farmers traditions are not respected due to the population displacement, the destruction of their encestral houses and tombs,,...

⋄ Abusive intensive exploitation of the mineral extraction which necessites a large amount of water and will dry up the water table.

⋄ Delicate mineral storage which should respect (in practice but not only on the specifications) the security norm about the emission of the radioactivity by the minerals released on the surface.

⋄ ...

4. Expected and Promised Positive Impacts of the Project

From the promise of the Company Energy Fuels Inc, the project will be expected to enhance the economy of the Region for creating local jobs and infrastrucutres in this Region. The Project will generate economical development of the Region by the money shuffling implied. However, some criticisms are raised by different assoications and personalities :

⋄ Local jobs for native are at low level as the Government recognizes that there are no local specialized Technicians and Engenieers who can work in the project.

⋄ Infrastructures like roads are primarily dedicated to the trucks and engines of the Company for tranporting materials and minerals but not for the whole population. Note that the huge traffic of trucks will certainly affect the environment.

⋄ There is not often an equal balance between the income of the Company and the return to the population and to the country due to the bad negociation when establishing the exploitation contract. Instead, the government administrators would have taken their personal benefit through some commissions when signing a such unbalanced contract.

⋄ ...

5. Conclusions

Weighing the pros and cons of this project, one can conclude that authorizing this project is very risky for the future of the population health and for the biodiversity and eco-systems. A priori, despite of the government expectations and company promise, this project cannot bring a global noticeable benefit and development for this South West Region of Madagascar. It will only favours a minority of official administrators, some government members and subcontracting companies !

References

[1] V.G Randrianandrasana, part of a PhD thesis presented at the conference HEP-MAD17, http://www.slac.stanford.edu/econf/C170921 (2017).

[2] J. Rajaobelison, INSTN (private email communication), October 2024.

[3] Z. Randriamaro, CRAAD-OI, These Proceedings.

[4] M. Rakotondrainibe, Collectif TANY, These Proceedings.

The Ranobe Base Toliara Project winners and loosers *

M. Rakotondrainibe

President of the Collectif pour la défense des terres malgaches – TANY
Boîte 89 - 8, rue du général Renault, Paris, 75011, France

Abstract

Madagascar's Minister of Mines announced that the suspension of activities at the Base Toliara project, in place since 2019, would be lifted in 2025. The American company Energy Fuels, which bought the Australian parent company Base Resources in 2024, is undoubtedly one of the winners. Will the Malagasy government, whose officials are currently negotiating tax matters with Energy Fuels, sell off rare non-renewable resources? Will the local communities, repressed by police because of their opposition to the reopening, lose out?

Keywords : mining - local community rights - land – water - tax –

---

1. Introduction

The Base Toliara mining concession covers an area of 40,753 hectaresa, located in the southwestern Atsimo-Andrefana region of Madagascar. As its size suggests, this concession includes 12 communes with an estimated population of over 200,000. The 3,200 hectare Ranobe deposit is located 45 kilometers north of the regional capital and the port of Toliara. This main mine is due to enter its exploitation phase and will have an average annual production of 780,000 tonnes of ilmenite, 53,000 tonnes of zircon and 7,000 tonnes of rutile. The Base Toliara project calls for the construction of on-site processing facilities, a power plant and a 55 kilometer haul road to link the mine to the existing port of Toliara.

The Base Toliara project was the subsidiary of the Australian company Toliara Sands that obtained a concession in 2012, and was purchased by Base Resources in 2018. This latter merged with the US-Canadian company Energy Fuels in 2024.

---

*Talk given at the 14th International Conference in High-Energy Physics – HEPMAD24, 21-26th October 2024, Antananarivo – Madagascar.

Email address : mamyrakoton@gmail.com

Created in late 2008 in France “to support the farmers and citizens of Madagascar in the defense of their land and natural resources”, the Collectif pour la défense des terres malgaches TANY has been following the project since 2014. To publicize the struggle of local communities opposing the project, it organized a conference with the Ma.Zo.To association in Paris in 2015 and appeared on TV5Monde in 2018.

While the Council of Ministers suspended the Base Toliara operation in 2019, - due to « disagreements between the company and the local community », the authorities have announced that Base Toliara will resume operations in 2025. The Collectif TANY continues to call for the closure of the Base Toliara project, as it considers that the Malagasy people will lose in this deal.

But let’s start speaking of the winners.

2. The winners

2.1. *The first winners are the mining companies and their shareholders*

The US-Canadian company Energy Fuels claims to become a world leader in the production of critical minerals notably uranium, rare earth elements and heavy mineral sands, thanks to its recent acquisitions in Madagascar, Brazil and Australia. This future world leader often write “at lower cost” when referring to products coming from Madagascar.

The shareholders of the Australian company Base Resources, which has just been bought by the American company Energy Fuels are also winners, as well as Base Toliara, the current local management of which will be entirely maintained by Energy Fuels. It will be very interesting to find out who the current local Base Toliara shareholders and associates are, following the suspension of operations and the ongoing negotiations between Energy Fuels and Malagasy decision-makers.

2.2. *Base Toliara's subcontractors and suppliers*

The Base Toliara's subcontractors and suppliers will see their business flourish.

2.3. *The foreign and Malagasy employees*

The number of jobs created remains a controversial subject: a promotional video made by Base Toliara broadcast by TVPlus in early October 2024 mentioned that the operation would last 38 years and that the number of jobs would be 4,350: 1,876 direct and 2,474 indirect. Knowing that the number of jobs generated by all mining operations is high during the construction phase and then drops considerably during subsequent phases, Base Toliara's promotional video did not specify the number of jobs envisaged before and then after the two-year construction phase.

As far as Malagasy employees hired, we don't know how many are expected. If all of them are recruited, we know there were 24 Malagasies were trained in Kenya for two years. Most of the jobs proposed to Malagasy workers are - for menial jobs such as drivers, janitors and cleaners.

The important question is : What percentage of the salaries paid will go to Malagasy employees?

2.4. *The countries of origin of the foreign companies will be winners,*

Especially the USA and Canada will benefit from the mining production: Energy Fuels wrote that their "supply of critical minerals will be secured" for decades to come, thanks to ilmenite, titanium, rutile, zircon and rare earths sourced (in part) from Madagascar. Subject to further permits and development, these heavy mineral sand projects should also produce a valuable by-product of monazite sand, which is one of the best sources of "magnetized" rare earths used in electric vehicles, plug-in hybrids, direct-drive wind power and other technologies. The production of rare earth oxides at the plant should be low-cost and globally competitive.

2.5. *There will be temporary winners*

The workers who will be recruited and employed only during the construction phase will be temporary winners.

Another group could be among the temporary winners: the Malagasy notables and local authorities, who for years have promoted, demanded and acted energetically for the resumption of Base Toliara's activities. Will they really be among the winners by receiving a permanent job or some kind of reward, or will they be "thanked" and forgotten when the continuation of Energy Fuels' activities is definitively assured?

Unfortunately, as in all mining operations, there will be losers if the Base Toliara project is resumed.

3. The losers

3.1. *The neighboring local communities, who will lose their land,*

their livelihood and their living space, are the main losers who will be the most affected. For the Collectif TANY, the losses that will be suffered by local communities are not only social, but also economic, since peasant farmers, livestock herders and fishermen make up three-quarters of the population in the area where the Base Toliara project is located. These farmers use the land as their main means of production to feed their families and as an essential source of income. What's more, when they're not being ruined by projects designed to develop and enrich other segments of the population, farmers are also investors. They sell their surplus produce on local markets, ensuring food security for the Malagasy population. Some export part of their production.

The Base Toliara promotional video mentions that local residents will receive compensation in accordance with the law. We would like to make a few comments on this subject. The 2019 pre-feasibility study for the Base Toliara project [1] mentioned the Malagasy laws with which the project would comply. Civil society organizations had protested vigorously because the law 2006-031 on untitled private property (which has since been revised), was not included in the list. The majority of community members who do not have title to their land, due to the prohibitive cost and difficulty of obtaining this document, run a high risk of not being compensated.

The amount of these compensations is a real subject for further study in the case of Base Toliara because a documentary on the mining company Rio Tinto QMM entitled "Je veux ma part de terre" (I want my share of the land) [2] shows people testify about insignificant compensations, some of them received ariary zato (100 ariary) per m^2, and fought for several years to obtain more.

The video claims to devote several million dollars to Base Toliara suppliers, including farmers, herders and fishers. But the list of suppliers is very long, and the farmers impacted by the project will certainly not benefit, especially if they no longer have their land.

A recurrent discussion occurs on expropriation for public use. We believe that the 45-kilometre road and the Andaboy port should not lead to expropriation for public use and evictions, as they will be reserved for the mining company's cars, trucks and boats exclusively.

3.2. *The neighboring local communities are likely to face serious problems of water unavailability and pollution*

In April 2023, hydrologists told the press that Base Toliara's water requirements would be 500 m^3 per hour per day, i.e. 4,380 million m^3 per year, and explained that this would not affect the water table used by local residents. In the current context of climate change, drought in the South and lack of water control worldwide, we remain highly skeptical.

Wastewater management is a crucial issue, and once again the experience of Rio Tinto QMM in Fort-Dauphin, which extracts and processes mineralized sands identical to those at Base Toliara, should serve as a lesson. Dozens of dead fish were found in a river near the company's site a few years ago, and since March 2024, the UK-based law firm Leigh Day has been taking legal action on behalf of villagers who accuse Rio Tinto of contaminating the streams and lakes they use for domestic purposes with high and harmful levels of uranium and lead, posing a serious risk to human health [3].

3.3. *The neighboring local communities and the workers at the mining project are exposed to radioactivity*

with risks of contracting serious illnesses and health problems that could also affect their offspring, such as birth defects. After the Base Resources' pre-feasibility study for the Ranobe site, published in March 2019 [1], the Definite Feasibility Study dated September 2021 [4] confirms that the zircon at the mine site contains high levels of uranium and thorium. The levels are even higher than the industrial standard, preventing the company from exporting the products to the USA and Japan. The monazite pre-feasibility study [5] revealed that the radioactivity of monazite is grade seven and should not be transported in the same vessels as other products.

3.4. *The Malagasy State is likely to be among the losers.*

Energy Fuels states that it is currently conducting tax negotiations with the Malagasy government. They request to benefit from the law on major mining investments whose main criterion is an investment amount in excess of 50 billion ariary (around 10 million Euros at the current exchange rate)

According to the document entitled « Madagascar, l'île aux trésors » (Madagascar, the treasure island) [6] inside the Economic Development Board of Madagascar web site, the tax advantages available to the mining company will be : Temporary exemption of the Minimum of Income Tax on Companies Perception, application of reduced rates : (Income Tax on Companies 25%, Transferable Capital Income Tax (IRCM), Inclusive Tax on Transfers (TFT), and to Developed Property Tax (IFPB) as well as to the IFPB supplementary tax (TAFPB), and additional IBS deductions .

The other privileges and advantages mentioned at the same place are : Freedom to convert at market rates ; duty-free importation of materials, goods and equipment; free importation of materials, goods and equipment; temporary admission under suspension of duties and taxes for equipment on the generic list; - possibility of setting up a special office for import and export operations. Concerning legal aspects: - for processed mining products, a 50% discount on their value for calculating the mining royalty base;- guarantee of property rights; - guarantee against expropriation or nationalization; and guaranteed free circulation and marketing of mining products".

Energy Fuels has a good chance of getting it all because of inequal power relationships.

In fact, the entire Malagasy population will be negatively impacted due to the loss of earnings and because of the direct and collateral consequences mentioned above.

3.5. *The Mikea Indigenous people in south-west Madagascar*, whose forest is being progressively eroded by the Base Toliara project, will be in danger of disappearing.

3.6. *The future Malagasy generations* will have important land issues.

If the mining company leaves and gives the land back at the end of operations, the land will be polluted and unfertile.

If the long term lease is constantly renewed and lasts *ad vitam eternam*, as Malagasy law allows, the riverside dwellers' descendants will have no land to inherit from their parents and ancestors, and no territory at all. The descendants of local residents will have no ancestral lands or territory to call their own.

3.7. *Indigenous peoples living near the White Mesa uranium processing area in Utah, USA*,

where the American company plans to send monazite from Toliara, will be among the losers, even though they are already struggling to cope with the impact of uranium processing on their natural resources (water...),

Conclusion :

At a time when the Malagasy people in general, and the communities of the Atsimo-Andrefana region in particular, are among the poorest on the planet, will the decisions of Madagascar's leaders be geared towards enriching Energy Fuels & C°, destroying the livelihoods of the people living near the mine site and violating their rights?

Will the decisions taken by the country's leaders further confirm the title of the 2012 document by Friends of the Earth, which stated that "Madagascar" was "the new Eldorado for mining and oil companies" but the destroyer of the lives of local communities and inhabitants?

Given the significant risks and dangers posed by the radioactivity of mining products to local communities'health and even their lives,

Given the lack of consideration for the fundamental rights of the inhabitants affected by the project, in particular their land rights and their right to food,

a resumption of Base Toliara's activities will impoverish and endanger the communities affected by this project, and exacerbate social injustice in Madagascar. So, the Collectif TANY continues to call for the definitive closure of the Base Toliara project.

References

[1] Base Resources, Toliara Project Pre-Feasibility study. Summary outcomes, (21 March 2019).

[2] F. Lambolez et J-M Pernelle, documentaire Je veux ma part de terre - Madagascar, (2012)

[3] The Intercept, Rio Tinto's Madagascar Mine Promised Prosperity. It Tainted a Community.

[4] Base Resources, Toliara Project Definite Feasibility Study 2. Summary Outcomes, (27 September 2021)

[5] Base Resources, Toliara Monazite Project Pre-Feasibility Study, (14 December 2023)

[6] Economic Development Board of Madagascar, https://edbm.mg/informations-economiques_secteurs_mines-madagascar-l-ile-aux-tresors/

Social and Ecological Costs of the Base Toliara project

Zo Randriamaro
Coordinator, Centre de Recherches et d'Appui pour les Alternatives de Développement – Océan Indien
Lot VF 60, Mahamasina Nord, Madagascar.

ABSTRACT

The Base Toliara project aims to become a major global source of heavy mineral sands and rare earth oxides by mining monazite in Madagascar. However, the project poses significant social and ecological costs. It will displace thousands of farmers and fisherfolks, leading to loss of livelihoods and access to vital natural resources without commensurate compensation. The project will also cause severe ecological degradation, including air and noise pollution, water depletion, and loss of biodiversity, impacting local communities' economic well-being. Additionally, it will disrupt cultural and religious practices, exacerbating social conflicts and resistance movements. Health risks from pollution and radioactivity are a major concern, as local communities may not afford medical treatment. The project will also contribute to climate change through greenhouse gas emissions from mining operations and diesel-powered generators. Overall, the negative impacts far outweigh the project's economic benefits, while generating social and ecological costs that are entirely externalized on local communities and the environment.

Keywords: mining – economic well-being – economic costs - external costs – ecological costs – social costs.

1. Introduction

In December 2023, the Australian mining company Base Resources announced its intention to make of its Base Toliara project one of the world's largest and most competitive sources of rare earth oxides by mining the monazite contained in the project's mineral sands. The acquisition of this integrated mineral sands and monazite project by US-based Energy Fuels took place in April 2024, as it reached the end of its exploration phase.

It is critically important to situate the Base Toliara project within its social and ecological context, whereby it is implemented within a Biodiversity Hotspot and one of the ecoregions of the Madagascar and Indian Ocean Islands supporting a rich fauna and flora with high levels of endemism. It is located in a dry, sub-arid ecological transition zone surrounded by protected areas (Huff, 2016) in one of the Districts most affected by poverty (81%) and drought in Madagascar, with a natural water deficit that results in a propensity for salinization of water and soil.

As such, the Base Toliara project will impose external costs that would be borne by the affected communities and future generations who are totally ecosystem-dependent.
These external costs are defined as those economic costs that result from an event or impact related to the Base Toliara project and are borne by a family, community, or generation that does not have any part in the event or impact, or otherwise directly benefit

from it (Niemi, 2015). The external costs further discussed below might materialize in many ways resulting from the project's activities and analysed in recent studies (Bisht and Martinez-Alier, 2022; Zaehringer et al, 2024).

2. External costs from physical and economic displacement of affected communities

Base Toliara will benefit from compulsory land acquisition through the Declaration of Public Utility (DUP), which will result in the expropriation from land of the communities of Ranobe, Benetse, Tsiafanoka and other communes affected by the construction of the project's mineral road and jetty.

Thus, thousands of farmers and fishermen will lose their land and livelihoods, while the planned workforce for the project is only around 400-535 employees (Base Resources, 2024). Moreover, they will suffer from the resulting social/land conflicts, loss of use of natural resources, and reduced access to water.

This expropriation process will also entail the relocation of tombs – which is traditionally forbidden in the local customs – as well as the loss of access to cultural and religious sites.

3. External costs from the degradation of ecosystems

External costs from ecological degradation by the Base Toliara project derive from its impacts on economic activity and on the economic well-being of society, with direct and indirect economic costs (Niemi, 2015). For instance, the risks of illness and death imposed on a community exposed to radioactivity entail a direct economic cost, as the fear of illness and death resulting from these risks imposed on the population constitute a reduction in economic well-being, i.e., an economic cost. Indirect economic costs will materialize as the project's activities – for instance the clearing of land - degrades an ecosystem's production of goods and services and this reduction decreases the economic well-being of the population that otherwise would have enjoyed these goods and services (Niemi, 2015).

Critical external costs linked to the degradation of ecosystems include air pollution by dust and radiation; noise pollution resulting from the transport of project products; and, potential costs from wastewater management.

3.1. External costs from climate change

The degradation of ecosystems brings about external costs from climate change, including the effects of global warming of 2-3°C typically linked to mining operations. The electrical energy needed to operate the Base Toliara mine will be supplied by diesel-powered generators, which will contribute to significant greenhouse gas emissions.

The traditional and sacred sites also constitute an important part of the cultural heritage and rights impacted by the so-called non-economic losses and damage from climate change.

3.2. External costs from the reduction in the quantity or value of goods and services available to communities and future generations

Most importantly, the loss of the Toliara II area's ecosystems and biodiversity will not only reduce tourism and associated economic benefits, but also the quantity or value of goods and ecosystem services available to local families, communities, and future generations.

Phase 1 of the Base Toliara project will last 21 years and covers a total area of around 455 hectares that will be cleared (ONE, 2014: 134), resulting in the loss of arable land, vegetation and biodiversity (particularly fauna and flora), including the baobabs endemic to the region. In addition, the reduced forest cover could lead to less rainfall in the region.

The planned sediment extraction and treatment processes will use a lot of water. Water requirements for dry mining and operating processes are estimated at 560 m3/h (just under 13,500 m3/day) for the duration of the project, of which approximately 90% will be extracted from boreholes (Base Resources, 2023). In a naturally semi-arid region, this is most likely to lower the groundwater table.

Research conducted in the villages surrounding the Base Toliara mine showed that biodiversity loss as a result of ecological degradation or changes had significant negative livelihood impacts, notably a loss of access to natural forests resources. About one-third of households located around the Ranobe site had reduced access to the sea or a river for fishing. The availability of fisheries resources, such as lobsters, had also deteriorated for almost all respondents (Zaehringer et al, 2024).

Farmers reported erosion and degradation of soils, while households observed a decrease in crop productivity, mostly attributed to a decrease in water availability. Most of those who did not feel safe were located around the Ranobe mining site and linked this impact to drought (Zaehringer et al, 2024).

5. External social costs

5.1. External Costs from Impairment of Cultural and Social Capital

The perceived negative ecological impacts of the mining project have also led to an ecological distribution conflict (EDCs), i.e. a community resistance movement against the mining project in Toliara II.

In Madagascar, land rights are intricately tied to cultural practices and possession of land itself is based upon the location of tombs of ancestors. People involved in the resistance movement state that "land cannot be viewed as a solely economic asset; it is also an existential anchor to past, present and future generations" (Seagle 2012: 455).

Local ecosystems that will be destroyed include several cultural and religious sites bearing witness to the region's history, and are also the guarantors of the customs and traditions still jealously guarded by the local population. As such, these have intrinsic archeological values, sacredness, or other cultural significance, which cannot be easily amended or compensated for through waged jobs or money indemnities.

Other social impacts of the extractive project, which require special attention, have been the subversion of the consent rights of local communities, and the violent repression against local communities and resisting actors.

5.2. External costs from loss of security, social harmony, quality of life, and public health issues

There had been serious friction between the Base Toliara project management and local communities, which peaked in April 2019, when forty people allegedly vandalized the mine's exploration campsite (Vyawahare, 2019) as it appeared that the mining company was ready to start operations.

These tensions around processes related to planning and implementing the Base Toliara project are consistent with global observations of increasing social resistance to sand mining projects due to their known ecologically destructive nature (Bisht and Martinez-Alier, 2023).

Research findings show that for 16% of households, social relations and opportunities to participate in social activities had deteriorated, mainly due to less unity and to people helping and respecting each other less because of the divergence of opinion about the mining project (Zaehringer et al., 2024).

Most households in the village of Benetse near the Ranobe site, thought that the Base Toliara project played a role in these changes, mainly because there was less unity in the village, people respected each other less, and there was a drought (Zaehringer et al., 2024).

Over the past 20 years, being healthy had become more difficult for 53% of households. When asked directly, almost one-quarter of households perceived mainly negative impacts of the project on their health, and related it to droughts (Zaehringer et al., 2024).

6. External Costs from Increases in Risk and Uncertainty

Since the announcement of the monazite project, public health concerns focus on the expected effects of pollution and radioactivity from the project, and on the costs that would be incurred by the affected poor communities who cannot afford medical treatment.

More generally, the Base Toliara project will impose external costs by increasing the risk that families and communities will be harmed to the extent that the project creates the risk of a major spill of contaminated water, radioactive elements or other hazardous material.

Furthermore, such a risk likely would increase anxiety of all affected individuals and communities, and reduce the value of residential, commercial, and public property.

7. Conclusions

The research results show that people living around the site of the Base Toliara project already perceive negative impacts during its exploration phase, even as the mining operations have not yet started (Zaehringer et al., 2024).

These negative ecological and socioeconomic impacts have been cost-shifted onto local communities. Moreover, the damage is often asymmetrically distributed even within affected communities, particularly with respect to gender, with women bearing disproportionately larger burdens (Bisht and Martinez-Alier, 2023). Thus, all these processes result in reinforcing societal inequities and injustices.

There are clearly multi-dimensional implications related to socioeconomic and ecological cost-shifting in terms of justice, equity, cultural heritage and values, and access to ecosystems that companies engaging in such resource extraction must be accountable for. In total, these external socioeconomic and ecological costs outweigh by far the alleged benefits from the Base Toliara integrated mineral sands and monazite project.

The empirical insights into the diverse and wide-ranging impacts of the Base Toliara project second the conclusions of other researchers about broader issues regarding the contribution of large scale extractive industries (LEIs) to sustainable development, that this kind of mining project directly impedes the achievement of at least seven (7) Sustainable Development Goals (SDGs), and that their expected contribution to economic growth cannot be considered 'sustainable' in the current circumstances. (Zaehringer et al, 2024).

References

Base Resources, 2023. Toliara Monazite Project Pre-Feasibility Study. https://mb.cision.com/Public/22548/3893356/89a168c3d231921e.pdf (accessed 15/10/2024)

Bisht, A., Martinez-Alier, J., 2023. Coastal sand mining of heavy mineral sands: contestations, resistance, and ecological distribution conflicts at HMS extraction frontiers across the world. J. Ind. Ecol. 27, 238–253. https://doi.org/10.1111/jiec.13358 (accessed 15/10/2024)

Niemi, E., 2015. Expected External Economic Costs of the Proposed Commercial Port at Lamu, Kenya. Report on behalf of Natural Justice and Save Lamu.

Office National de l'Environnement (ONE), 2014. Plan de Gestion Environnementale, Projet d'exploitation d'ilmenite, de Rutile et de Zircon de la Société Toliara Sands - Pe N° 37242. Communes rurales de Tsianisiha – Ankilimalinike – Belalanda – Maromiandra et Commune urbaine de Toliara - District de Toliara II - Région Atsimo Andrefana.

Seagle, C., 2012. Inverting the impacts: mining, conservation and sustainability claims near the Rio Tinto/QMM ilmenite mine in Southeast Madagascar. J. Peasant Stud. 39, 447–477.

Vyawahare, M., 2019. Madagascar Suspends Activities at Controversial Base Toliara Mine [WWW Document]. Mongabay Environmental News. URL. https://news.mongabay.com/2019/11/madagascar-suspends-activities-at-controversial-base-toliara-mine/ (accessed 7.05.23).

Zaehringer, J.G. et al., 2024. How are large-scale extractive industries affecting progress toward the sustainable development goals in Madagascar? Perceived social-ecological impacts of mining investments, Current Research in Environmental Sustainability, Volume 8, 2024,100257. https://doi.org/10.1016/j.crsust.2024.100257 (accessed 15/10/2024)

Zaehringer, J.G. et al., 2023. Governance processes and sustainability impacts of the extractive industries: Generating transformation knowledge in the biodiversity hotspot of Madagascar. Final scientific report. https://snis.ch/wp-content/uploads/2020/01/Final-scientific-report_including-executive-summary.pdf (accessed 15/10/2024)

Open Discussions by the Participants

H.T. Rakotondramanana (Dean of the Faculty of Sciences, Antananarivo, MG)

M. Andrianainarivelo (Vice-Dean of the Faculty of Sciences, Antananarivo, MG)

M.H. Rabeony & V. Andriamanantenasoa (CRAAD-OI, Antananarivo, MG)

V.G. Randrianandrasana (Univ. Tulear, MG)

D. Randimbimiariso (Parti Otrikafo, Antananarivo, MG)

M. Rakotondrainy (Entrepreneur, Antananarivo, MG).

M. Rakotoanosy (Univ. Antananarivo, MG)

Part 3

QCD and Hadron Physics

Chairs: S. Narison, C. Ratsifaritana

Recent highlights from the Belle II experiment

Shu-Ping Lin

Department of Physics and Astronomy, University of Padova and INFN,
Via Marzolo 8, Padova 35131, Italy

Received 25 March 2025
Accepted 31 July 2025
Published 1 November 2025

The Belle II experiment operates at the SuperKEKB asymmetric-energy e^+e^- collider. During the Run 1 data taking, we have collected an integrated luminosity of 424 fb^{-1} of collision data at the energy near the $\Upsilon(4S)$ resonance. We present highlights of recent Belle II results on measurements of rare B decays, tests of lepton flavor universality, and measurements that contribute to the determination of the Cabibbo–Kobayashi–Maskawa unitarity triangle and the muon anomalous magnetic moment.

Keywords: B decays; CKM unitary triangle; muon g-2.

1. Introduction

The Belle II experiment[1] is located at the asymmetric-energy SuperKEKB[2] accelerator in Tsukuba, Japan. It operates near the $\Upsilon(4S)$ resonance, allowing for production of B meson pairs at threshold. SuperKEKB was designed to achieve an instantaneous luminosity that is significantly higher than its predecessor, KEKB,[3] by implementing the nano-beam scheme. It has achieved instantaneous luminosity of $4.7 \times 10^{34}\ \mathrm{cm}^{-2}\,\mathrm{s}^{-1}$, setting a new world record.

The Belle II detector consists of multiple sub-detectors. At the innermost, there are two layers of silicon pixel detector and four layers of silicon strip detector. Along with the central drift chamber, these sub-detectors reconstruct the tracks and vertices of charged particles. Particle identification is provided by a time-of-propagation counter in the barrel region and an aerogel-based ring-imaging Cherenkov counter in the forward region. An Electromagnetic Calorimeter (ECL) based on CsI (Tl) crystals provides energy and timing information for photons and electrons. All the sub-detectors described so far operate within a 1.5 T magnetic field generated by a superconducting solenoid magnet. The K_L^0 and muon detector are embedded in the flux return of the solenoid.

Belle II has several advantages compared to experiments conducted on proton–proton machines. The well-defined initial state and clean environment allow for precise measurements. The hermetic detector structure is ideal for decays to neutral or invisible particles. The coherent B meson pair production significantly enhances the flavor tagging[4] efficiency, and at e^+e^- colliders it is possible to perform full event interpretation.[5]

During the data-taking at the SuperKEKB collider in 2019–2022 (Run 1), Belle II collected an integrated luminosity of $424\,\text{fb}^{-1}$ of collision data, including $362\,\text{fb}^{-1}$ at the $\Upsilon(4S)$ resonance. We present selected recent results from the Belle II experiment, including measurements of rare B decays, tests of Lepton Flavor Universality (LFU), and measurements that contribute to the determination of the Cabibbo–Kobayashi–Maskawa (CKM) unitarity triangle[6,7] and the muon anomalous magnetic moment.

2. Rare B Decays

2.1. *Evidence for $B^+ \to K^+\nu\bar{\nu}$ decays*

The decay $B^+ \to K^+\nu\bar{\nu}$ is a Flavor-Changing-Neutral-Current (FCNC) process that is suppressed in the Standard Model (SM) at tree level due to CKM and Glashow-Iliopoulos–Maiani suppressions.[8] The SM prediction for the branching fraction is $(5.6 \pm 0.4) \times 10^{-6}$.[9] Deviation from this prediction may indicate contributions from nonSM physics. The branching fraction can be significantly modified in models that predict nonSM particles, or the B meson could be decaying to a kaon and an invisible particle, such as a dark matter candidate.

The full Belle II Run 1 dataset ($362\,\text{fb}^{-1}$) collected at the $\Upsilon(4S)$ resonance is used for this analysis. Two different and largely independent approaches have been used: One using the Inclusive B Tagging Method (ITA) and the other using the Hadronic B TAgging method (HTA).[5] The two samples have a very small overlap, which makes the combination of the results quite straightforward. The ITA achieves better sensitivity, and the HTA serves as a consistency check, providing a 10% increase in precision in the final combination. The missing energy is reconstructed in the form of the mass squared ($q^2_{\text{rec}} = s/(4c^4) + M_K^2 - \sqrt{s}E^*_K/c^4$) of the neutrino pair, where M_K is the known mass of K^+ mesons and E^*_K is the reconstructed energy of the kaon in the center-of-mass frame. The B meson is assumed to be at rest in the center-of-mass frame for the calculation of q^2_{rec}.

For ITA, two consecutive multivariate classifiers are trained using signal kaon, event shape and rest-of-event information. The first classifier serves as a first-level filter and the second classifier $\eta(\text{BDT}_2)$ is used for final event selection. The signal yield is determined in bins of q^2_{rec} and $\eta(\text{BDT}_2)$. For HTA, only one classifier $\eta(\text{BDTh})$ is trained. We report the measured branching fractions[10]

$$\mathcal{B}_{\text{ITA}}(B^+ \to K^+\nu\bar{\nu}) = (2.7 \pm 0.5 \pm 0.5) \times 10^{-5},$$
$$\mathcal{B}_{\text{HTA}}(B^+ \to K^+\nu\bar{\nu}) = (1.1^{+0.9+0.8}_{-0.8-0.5}) \times 10^{-5}$$

and the combined result

$$\mathcal{B}(B^+ \to K^+\nu\bar{\nu}) = (2.3 \pm 0.5 \pm 0.5) \times 10^{-5}.$$

The ITA and HTA results are consistent within 1.2σ. We observe a 3.5σ significance and a 2.7σ deviation from the SM prediction. This is the first evidence for $B^+ \to K^+\nu\bar{\nu}$ decays.

2.2. *Search for $B^0 \to K^{*0}\tau^+\tau^-$ decays*

The decay $B^0 \to K^{*0}\tau^+\tau^-$ also proceeds through an FCNC process. The SM prediction of this branching fraction is $(0.98 \pm 0.10) \times 10^{-7}$.[11,12] Non-SM theories accommodating $b \to c\tau^-\bar{\nu}$ anomalies[13] predict enhancements of the branching fraction by several orders of magnitude for processes with a τ pair in the final state. In some scenarios,[14] the leading new physics couplings involve the third-fermion generation, making this channel a better probe compared to final states with a pair of electrons or muons. Milder enhancements are also foreseen in models that explain the $B^+ \to K^+\nu\bar{\nu}$ excess.[12]

The full Run 1 $\Upsilon(4S)$ dataset ($362\,\text{fb}^{-1}$) is used for this analysis. For this channel, there could be up to four neutrinos in the final state, and the missing momentum information is deduced using hadronic B-tagging.[5] Reconstruction is challenging because the signal component does not peak in any kinematic observable, and the K^* mesons typically have low momenta due to phase space. Four signal categories can be defined based on how the two τ leptons are reconstructed. A multivariate classifier is trained to separate signal and background, and the branching fraction is extracted from a fit to this classifier, simultaneously over all four signal categories. No evidence for a signal is observed. We obtain an upper limit on the branching fraction

$$\mathcal{B}(B^0 \to K^{*0}\tau^+\tau^-) < 1.73 \times 10^{-3},$$

at 90% confidence level. This result represents the most stringent result to date in general for $b \to s\tau^+\tau^-$ transitions. It achieves two times better precision than Belle, using a dataset that is only about half the size.

2.3. *Search for $B^0 \to \gamma\gamma$ decays*

In the SM, there is no tree-level interaction between the b and d quarks. The double radiative decay $B^0 \to \gamma\gamma$ proceeds via an FCNC transition involving electroweak loop amplitudes. The SM prediction of the branching fraction of $B^0 \to \gamma\gamma$ decays is $(1.4^{+1.4}_{-0.8}) \times 10^{-8}$,[15] which is difficult to calculate due to long-distance contributions. This channel is sensitive to contributions of non-SM particles in the loop.[16–18]

This channel is characterized by two nearly back-to-back highly energetic photons. Two multivariate classifiers are trained in order to reduce the background. The first classifier vetoes photons coming from decays of π^0 and η. It is trained using the diphoton mass and ECL cluster shape variables. The second classifier is the Continuum Suppression (CS), which separates the background from $e^+e^- \to q\bar{q}$ processes.

The full Belle II Run 1 dataset ($362\,\mathrm{fb}^{-1}$) and the Belle data with ECL timing information ($694\,\mathrm{fb}^{-1}$) collected at the $\Upsilon(4S)$ resonance are used in this analysis. A simultaneous three-dimensional fit of ΔE, the beam constrained mass M_{bc}, and the CS output is performed to extract the branching fraction. We obtain an upper limit on the branching fraction[19] using the Belle and Belle II datasets

$$\mathcal{B}(B^0 \to \gamma\gamma) < 6.4 \times 10^{-8},$$

at 90% confidence level. The uncertainties of Belle and Belle II are comparable, and there is a five-times improvement over the previous upper limit.[20] The sensitivity of this analysis approaches the SM prediction.

3. Lepton Flavor Universality

In the SM, e, μ and τ leptons are predicted to have the same coupling strengths with the W bosons. This is referred to as the LFU. Semileptonic decays of the B meson allow for test of the LFU by measuring the ratio

$$\mathcal{R}(D^{(*)}) = \frac{\mathcal{B}(B \to D^{(*)}\tau^-\bar{\nu}_\tau)}{\mathcal{B}(B \to D^{(*)}\ell^-\bar{\nu}_\ell)}, \tag{1}$$

where $\ell = e$ or μ. The measurement of the ratio allows for many theoretical and experimental uncertainties to cancel, e.g. $|V_{cb}|$. The cancellation makes these measurements using stringent LFU tests. The theoretical predictions are $\mathcal{R}(D) = 0.298 \pm 0.004$ and $\mathcal{R}(D^*) = 0.254 \pm 0.005$.[21] There is a long-standing 3.3σ tension between world averages of experimental measurements and the SM prediction. This could indicate enhanced coupling of the b quark to the τ lepton, as predicted in some beyond SM scenarios.[22,23]

3.1. *Measurement of* $\mathcal{R}(D^*)$

The ratio of branching fractions is measured using $189\,\mathrm{fb}^{-1}$ of Belle II Run 1 data collected at the $\Upsilon(4S)$ resonance. Reconstructed D^* decays include $D^{*+} \to D^0\pi^+$, $D^{*+} \to D^+\pi^0$ and $D^{*0} \to D^0\pi^0$. Hadronic B tagging[5] is used in order to obtain information about the missing momentum, and signal τ is only reconstructed from leptonic decays. $\mathcal{R}(D^*)$ is extracted from a simultaneous 2D fit to the missing mass squared $M^2_{\mathrm{miss}} = (E^*_{\mathrm{beam}} - E^*_{D^*} - E^*_\ell)^2 - (-\vec{p}^{\,*}_{B_{\mathrm{tag}}} - \vec{p}^{\,*}_{D^*} - \vec{p}^{\,*}_\ell)^2$, and the residual ECL energy, which is the sum of the energies detected in the ECL that is not associated with the reconstructed $B\bar{B}$ pair. We obtain

$$\mathcal{R}(D^*) = 0.262\,^{+0.041+0.035}_{-0.039-0.032}.$$

The result[24] has a comparable precision to Belle[25] despite using a much smaller dataset, and is consistent with the current world average and the SM prediction.

3.2. *Measurement of* $\mathcal{R}(X_\tau/\ell)$

The inclusive branching fraction ratio $\mathcal{R}(X_{\tau/\ell})$ is given by

$$\mathcal{R}(X_{\tau/\ell}) = \frac{\mathcal{B}(\bar{B} \to X\tau^-\bar{\nu}_\tau)}{\mathcal{B}(\bar{B} \to X\ell^-\bar{\nu}_\ell)}, \tag{2}$$

where X is a generic hadronic final state originating from $b \to c\tau^-\bar{\nu}$ or $b \to u\tau^-\bar{\nu}$ decays. The ratio is measured using $189\,\text{fb}^{-1}$ of Belle II Run 1 data collected at the $\Upsilon(4S)$ resonance. Both D and D^* are incorporated, regardless of their subsequent decay modes, as well as contribution from unexplored semitauonic B decays. The inclusive ratio is based on different theoretical inputs than from the exclusive ratios $\mathcal{R}(D^{(*)})$,[26,27] therefore this is a statistically and theoretically distinct LFU test.

This analysis utilizes the hadronic B tagging.[5] We utilize only $\tau \to \ell\nu_\ell\bar{\nu_\tau}$ decays, reconstructing only the light lepton ℓ from the signal side, and the remaining particles are assigned to X. The result is extracted from a 2D maximum likelihood fit to the lepton momentum in the B rest frame and the missing mass squared. The electron and muon results are combined,

$$\mathcal{R}(X_{\tau/\ell}) = 0.228 \pm 0.016 \pm 0.036,$$

which is systematically limited due to the size of the control sample, and is in agreement with the SM prediction. It is also consistent with a hypothetically enhanced semitauonic branching fraction as indicated by the $\mathcal{R}(D^*)$ world averages.[21] The total correlation between $\mathcal{R}(X_{\tau/\ell})$ and the exclusive measurement of $\mathcal{R}(D^*)$ in Sec. 3.1 is estimated to be below 0.1, therefore $\mathcal{R}(X)$ is a largely independent probe of the $b \to c\tau^-\bar{\nu}$ anomaly.

4. CKM Measurements

Precise measurement of the CKM quark mixing parameters is crucial for understanding the quark mixing and CP violation in the SM. Measuring the CKM parameters at very high precisions constrains the unitarity triangle and tests the consistency of the SM. It is also a sensitive probe to nonSM physics.

4.1. *Determination of* ϕ_2

The angle $\phi_2 = \arg(-V_{td}V_{tb}^*/V_{ud}V_{ub}^*)$ is the least precisely known angle, where V_{ij} are elements of the CKM matrix. It can be accessed via $b \to u$ transitions. However, $b \to d$ loop contributions shift the observed ϕ_2 by $\Delta\phi_2$. The impact of hadronic uncertainties can be reduced exploiting isospin symmetry[28] via the combined information of branching fraction and CP asymmetry measurements of the full set of isospin-related $B \to \rho\rho$ decays, $B \to \pi\pi$ decays, or from the Dalitz analysis of $B \to \rho\pi$ decays. The current world average is $\phi_2^{\text{WA}} = (85.2^{+4.8}_{-4.3})^\circ$.[29] Belle II has the unique ability to measure all the relevant decay channels, many of which have already been performed.[30–33]

4.1.1. *Measurement of the branching fraction and $\mathcal{A}_{\rm CP}$ of $B^0 \to \pi^0\pi^0$*

The $B^0 \to \pi^0\pi^0$ channel is both CKM-suppressed and color-suppressed. Theoretical predictions for the branching fraction involve hadronic amplitudes and are therefore challenging.

The full Belle II Run 1 dataset (362 fb^{-1}) collected at the $\Upsilon(4S)$ resonance is used for this analysis. The final state of $B^0 \to \pi^0\pi^0$ decays consists of only photons, whose kinematics are measured less precisely than charged particles at Belle II. It is also affected by energy leakage and beam backgrounds. In this analysis, a dedicated classifier is trained using ECL information to suppress the background from fake photons. Another classifier is trained to suppress the continuum background, which is the dominant background component. The graph-neural-network-based flavor tagger[4] is used to determine the flavor of the B meson.

The branching fraction and direct CP asymmetry $\mathcal{A}_{\rm CP}$ are extracted from a simultaneous fit to M_{bc}, ΔE, the CS classifier, and the flavor tagger output (w). In contrast to other analyses, where the fit is performed in bins of w, it is included directly as one of the fit variables in this analysis. The fit configuration is data-driven in order to reduce systematic uncertainties. We obtain

$$\mathcal{B}(B^0 \to \pi^0\pi^0) = (1.26 \pm 0.20 \pm 0.12) \times 10^{-6},$$

$$\mathcal{A}_{\rm CP}(B^0 \to \pi^0\pi^0) = 0.06 \pm 0.30 \pm 0.05.$$

The result[34] is compatible with world averages, achieving the world best precision with the branching fraction, and a comparable precision to the world best with the CP asymmetry measurement.

4.1.2. *Time-dependent measurement of $B^0 \to \rho^+\rho^-$*

The decay $B^0 \to \rho^+\rho^-$ has a small contribution from loop amplitudes and gives the most stringent constraints on ϕ_2. An angular analysis is required as this is a pseudo-scalar to vector-vector decay. The final state consists of three possible helicity states: One longitudinally polarized state and two transversely polarized states. The polarization information can be accessed from the helicity angles.

The full Belle II Run 1 dataset (362 fb^{-1}) collected at the $\Upsilon(4S)$ resonance is used for this analysis. The CS of this analysis is trained with the TabNet algorithm.[35] In addition, there is a dedicated classifier to suppress fake photons from the subsequent decays of ρ mesons. The selections of this analysis is optimized with differential evolution,[36] iterating over multivariate trainings and selection optimization to reach the optimal performance.

To extract the branching fraction, longitudinal polarization f_L, and the CP asymmetries $\mathcal{S}_{\rm CP}$ and $\mathcal{C}_{CP}$, we perform fits to nine variables, ΔE, the ρ masses $m_{\pi^\pm\pi^0}$ and helicity angles $\cos\theta_{\rho_\pm}$, the transformed CS classifier $\mathcal{T}_C$, the decay time difference between the signal-side and tag-side B meson Δt, the flavor of the tag-side B meson q, and the flavor tagging quality r. Two maximum-likelihood fits are performed: A signal extraction fit for the branching fraction and f_L, using ΔE, $m_{\pi^\pm\pi^0}$, $\cos\theta_{\rho_\pm}$, and

$\mathcal{T}_C$, and then a time-dependent CP-asymmetry fit. $\mathcal{S}_{\rm CP}$ and $\mathcal{C}_{\rm CP}$ are determined from a fit to Δt, its uncertainty $\sigma_{\Delta t}$, and q in seven bins of r. Correlations between fitting variables are taken into account by modelling one variable in bins of another.

We obtain from the signal extraction fit

$$\mathcal{B}(B^0 \to \rho^+\rho^-) = (29.0^{+2.3}_{-2.2}\,^{+3.1}_{-3.0}) \times 10^{-6},$$
$$f_L(B^0 \to \rho^+\rho^-) = 0.921^{+0.024}_{-0.025}\,^{+0.017}_{-0.015},$$

with a statistical correlation coefficient of -0.11, and from the time-dependent CP-asymmetry fit

$$\mathcal{C}_{\rm CP}(B^0 \to \rho^+\rho^-) = -0.02 \pm 0.12^{+0.06}_{-0.05},$$
$$\mathcal{S}_{\rm CP}(B^0 \to \rho^+\rho^-) = -0.26 \pm 0.19 \pm 0.08.$$

The results[37] of the analysis are mostly statistically limited except for the branching fraction, and are consistent with previous measurements.

An isospin analysis is performed to determine ϕ_2, incorporating the world averages of all the $B \to \rho\rho$ measurements in addition to the result of this analysis. The difference in the branching fractions of $\Upsilon(4S)$ decay to B^+B^- or $B^0\bar{B}^0$ is also taken into account. We obtain

$$\phi_2 = (92.6^{+4.5}_{-4.8})^\circ.$$

The sensitivity of ϕ_2 is improved by 6% compared to the sensitivity without inputs from this measurement, and is limited by the precisions of the time-dependent CP asymmetries of $B^0 \to \rho^+\rho^-$ and $B^0 \to \rho^0\rho^0$ decays.

4.2. *Determination of* ϕ_3

The angle $\phi_3 = \arg(-V_{ud}V^*_{ub}/V_{cd}V^*_{cb})$ can be accessed by interfering $b \to c$ and $b \to u$ processes to the same final state. The processes are tree-level dominated, with no large contributions from physics beyond the SM. We present a first combination of all Belle and Belle II measurements, implementing different methods and with data samples of different sizes, as listed in Table 1. The sensitivity is mostly led by the BPGGSZ method.[38–40] The final combination[41] is

$$\phi_3 = (75.2 \pm 7.6)^\circ.$$

It is consistent with the current world average, $\phi_3^{\rm WA} = (66.4^{+2.8}_{-3.0})^\circ$.[29]

4.3. *Determination of* $|V_{ub}|$

The determination of $|V_{ub}|$ is important to constrain the CKM unitarity triangle. $|V_{ub}|$ can be precisely measured with semileptonic B decays. There is a long-standing tension between exclusive and inclusive determinations of about 2.5σ.[21] For the exclusive determination, a specific final state is reconstructed, and it is described theoretically using form factors. For the inclusive determination, no specific final

Table 1. Belle and Belle II measurements used for the combination of ϕ_3.

B decay	D decay	Method	Data set [fb^{-1}] (Belle + Belle II)	Ref.
$B^+ \to Dh^+$	$D \to K_S^0\pi^0, K^-K^+$	GLW	$711 + 189$	42
$B^+ \to Dh^+$	$D \to K^+\pi^-, K^+\pi^-\pi^0$	ADS	$711 + 0$	43 and 44
$B^+ \to Dh^+$	$D \to K_S^0K^-\pi^+$	GLS	$711 + 362$	45
$B^+ \to Dh^+$	$D \to K_S^0h^-h^+$	BPGGSZ	$711 + 128$	46
$B^+ \to Dh^+$	$D \to K_S^0\pi^-\pi^+\pi^0$	BPGGSZ	$711 + 0$	47
$B^+ \to D^*K^+$	$D^* \to D\pi^0$ $D \to K_S^0\pi^0, K_S^0\phi, K_S^0\omega, K^-K^+, \pi^-\pi^+$	GLW	$210 + 0$	48
$B^+ \to D^*K^+$	$D^* \to D\pi^0, D\gamma$ $D \to K_S^0\pi^-\pi^+$	BPGGSZ	$605 + 0$	49

state is reconstructed. Instead, the sum of all possible final states is analyzed. Theoretically, it is described with calculation of the total semileptonic decay rate. The current world average is $|V_{ub}|^{\mathrm{WA}} = (3.67 \pm 0.09 \pm 0.12) \times 10^{-3}$.[21]

We report a result using the exclusive final states $B^0 \to \pi^-\ell^+\nu_\ell$ and $B^+ \to \rho^0\ell^+\nu_\ell$. The full Belle II Run 1 dataset ($362\,\mathrm{fb}^{-1}$) collected at the $\Upsilon(4S)$ resonance is used. The sample is untagged, and the neutrino momentum is estimated from all reconstructed tracks and clusters. Continuum and $B\bar{B}$ background are suppressed using dedicated classifiers. The $B^0 \to \pi^-\ell^+\nu_\ell$ and $B^+ \to \rho^0\ell^+\nu_\ell$ signal yields are extracted from a simultaneous fit to binned distributions of M_{bc}, ΔE, and q^2. The measured branching fractions[50] are

$$\mathcal{B}(B^0 \to \pi^-\ell^+\nu_\ell) = (1.516 \pm 0.042 \pm 0.059) \times 10^{-4},$$

$$\mathcal{B}(B^+ \to \rho^0\ell^+\nu_\ell) = (1.625 \pm 0.079 \pm 0.180) \times 10^{-4},$$

where the dominant systematic uncertainty arises from the off-resonance sample size. We perform a simultaneous measurement of the differential branching fractions as a function of q^2, and $|V_{ub}|$ is extracted separately from each decay mode using χ^2 fits to the q^2 spectra, using constraints on the form factors from lattice QCD and Light-Cone Sum Rule (LCSR). The result[50] extracted from $B^0 \to \pi^-\ell^+\nu_\ell$ decays using LQCD constraints[51] is

$$|V_{ub}| = (3.93 \pm 0.09\ (\mathrm{stat.}) \pm 0.13\ (\mathrm{syst.}) \pm 0.19\ (\mathrm{theo.})) \times 10^{-3}.$$

Using additional constraints from LCSR,[52]

$$|V_{ub}| = (3.73 \pm 0.07\ (\mathrm{stat.}) \pm 0.07\ (\mathrm{syst.}) \pm 0.16\ (\mathrm{theo.})) \times 10^{-3}.$$

From $B^+ \to \rho^0\ell^+\nu_\ell$ decays, with constraints from LCSR,[53]

$$|V_{ub}| = (3.19 \pm 0.12\ (\mathrm{stat.}) \pm 0.18\ (\mathrm{syst.})\ \pm 0.26\ (\mathrm{theo.})) \times 10^{-3}.$$

The most precise result is from the $B^0 \to \pi^- \ell^+ \nu_\ell$ channel, where additional constraints from LCSR further reduce the uncertainties. The results are consistent with the world average and are limited by theoretical uncertainties.

5. Measurement of the $e^+e^- \to \pi^+\pi^-\pi^0$ Cross-Section

The hadronic cross-section for e^+e^- annihilation at $\sqrt{s} < 2$ GeV is an input to dispersion relations that predict a_μ, where the largest source of uncertainty arises from knowledge of the cross-section.

There is a 5σ discrepancy between the SM dispersive prediction and experimental measurements.[54] Recent predictions based on lattice QCD[55] show a 2 to 3σ difference from values based on dispersion relations. The difference from the measured muon anomalous magnetic moment is smaller for this calculation. There is a long-standing difference between the BaBar[56] and KLOE[57] measurements for the $\pi^+\pi^0$ final state, which contributes to the systematic uncertainty of a_μ. Therefore, additional experimental measurements are crucial in order to clarify the situation.

We use an e^+e^- data sample corresponding to 191 fb^{-1} of integrated luminosity, collected at or near the $\Upsilon(4S)$ resonance. This analysis implements the initial state radiation method.[58] This allows measurement of the cross-section as a function of the center-of-mass energy without having to directly vary the collision energy. The energy ranges from 620 MeV to 3.5 GeV. The signal is extracted by fitting the diphoton mass in each bin of the invariant mass of the three pions. The resulting contribution[60] at leading order in the HVP to the muon anomalous magnetic moment is

$$a_\mu^{3\pi} = (48.91 \pm 0.23 \pm 1.07) \times 10^{-10}.$$

The result is 2.5σ higher than the results from BaBar[56] or the global fit.[59] The dominant systematic uncertainty arises from efficiency corrections.

6. Summary

In summary, we highlight ten new results from the Belle II experiment. We have achieved precisions on par with Belle and BaBar results despite using a smaller dataset. This is only a small fraction of exciting new results from Belle II, and we are expecting a significant increase of data in the Run 2 data collection.

References

1. Belle II Collab. (T. Abe *et al.*), arXiv:1011.0352.
2. SuperKEKB (K. Akai, K. Furukawa and H. Koiso), *Nucl. Instrum. Methods Phys. Res. Sec. A* **907**, 188 (2018).
3. T. Abe *et al.*, *Prog. Theor. Exp. Phys.* **2013**, 03A001 (2013).
4. Belle II Collab. (I. Adachi *et al.*), *Phys. Rev. D* **110**, 012001 (2024).
5. T. Keck *et al.*, *Comp. Soft. Big Sci.* **3**, 6 (2019).
6. N. Cabibbo, *Phys. Rev. Lett.* **10**, 531 (1963).

7. M. Kobayashi and T. Maskawa, *Prog. Theor. Phys.* **49**, 652 (1973).
8. S. Glashow, J. Iliopoulos and L. Maiani, *Phys. Rev. D* **2**, 1285 (1970).
9. HPQCD Collab. (W. Parrott *et al.*), *Phys. Rev. D***107**, 014511 (2023); **107**, 119903(E) (2023).
10. Belle II Collab. (I. Adachi *et al.*), *Phys. Rev. D* **109**, 112006 (2024).
11. J. Hewett, *Phys. Rev. D* **53**, 4964 (1996).
12. B. Capdevila *et al.*, *Phys. Rev. Lett.* **120**, 181802 (2018).
13. L. Allwicher *et al.*, *Phys. Lett. B* **848**, 138411 (2024).
14. C. Cornella *et al.*, *J. High Energy Phys.* **2021**, 050 (2021).
15. Y.-L. Shen, Y.-M. Wang and Y.-B. Wei, *J. High Energy Phys.* **2020**, 169 (2020).
16. G. Devidze and G. Jibuti, *Phys. Lett. B* **429**, 48 (1998).
17. G. Devidze, A. Liparteliani and Ulf-G. Meiner, *Phys. Lett. B* **634**, 59 (2006).
18. J. Aranda, *et al.*, *Phys. Rev. D* **82**, 054002 (2010).
19. Belle II Collab. (I. Adachi *et al.*), *Phys. Rev. D* **110**, L031106 (2024).
20. BABAR Collab. (P. del Amo Sanchez *et al.*, *Phys. Rev. D* **83**, 032006 (2011).
21. HFLAV Collab. (Y. Amhis *et al.*), *Phys. Rev. D* **107**, 052008 (2023).
22. Y. Grossman and Z. Ligeti, *Phys. Lett. B* **332**, 373 (1994).
23. D. Beirevi *et al.*, *Phys. Rev. D* **98**, 055003 (2018).
24. Belle II Collab. (I. Adachi *et al.*), arXiv:2401.02840.
25. Belle Collab. (M. Huschle *et al.*), *Phys. Rev. D* **92**, 072014 (2015).
26. M. Rahimi and K. Vos, *J. High Energ. Phys.* **2022**, 7 (2022).
27. M. Freytsis, Z. Ligeti and J. Ruderman, *Phys. Rev. D* **92**, 054018 (2015).
28. M. Gronau and D. London, *Phys. Rev. Lett.* **65**, 3381 (1990).
29. Particle Data Group (S. Navas *et al.*), *Phys. Rev. D* **110**, 030001 (2024).
30. Belle II Collab. (F. Abudinn *et al.*), arXiv: 2208.03554.
31. Belle II Collab. (F. Abudinn *et al.*), arXiv:2206.12362.
32. Belle II Collab. (I. Adachi *et al.*), *Phys. Rev. D* **109**, 012001 (2024).
33. Belle II Collab. (F. Abudinn *et al.*), *Phys. Rev. D* **107**, 112009 (2023).
34. Belle II Collab. (I. Adachi *et al.*), arXiv:2412.14260.
35. S. Arik and T. Pfister, arXiv: 1908.07442.
36. R. Storn and K. Price, *J. Glob. Optim.* **11**, 341 (1997).
37. Belle II Collab. (I. Adachi *et al.*), arXiv: 2412.19624.
38. A. Giri *et al.*, *Phys. Rev. D* **68**, 054018 (2003).
39. A. Bondar and A. Poluektov, *Eur. Phys. J. C* **55**, 51 (2008).
40. A. Bondar and A. Poluektov, *Eur. Phys. J. C* **47**, 347 (2006).
41. Belle II Collab. (I. Adachi *et al.*), *J. High Energ. Phys.* **2024**, 143 (2024).
42. Belle and Belle II Collab. (I. Adachi *et al.*), *J. High Energy Phys.* **2024**, 212 (2024).
43. Belle Collab. (M. Nayak *et al.*), *Phys. Rev. D* **88**, 091104 (2013).
44. Belle Collab. (Y. Horii *et al.*), *Phys. Rev. Lett.* **106**, 231803 (2011).
45. Belle and Belle II Collab. (I. Adachi *et al.*), *J. High Energy Phys.* **2023**, 146 (2024).
46. Belle and Belle II Collab. (F. Abudinn *et al.*, *J. High Energ. Phys.* **2022**, 63 (2022).
47. Belle Collab. (P. Resmi *et al.*), *J. High Energy Phys.* **2019**, 10 (2019).
48. Belle Collab. (K. Abe *et al.*, *Phys. Rev. D* **73**, 051106 (2006).
49. Belle Collab. (A. Poluektov *et al.*), *Phys. Rev. D* **81**, 112002 (2010).
50. Belle II Collab. (I. Adachi *et al.*), arXiv:2407.17403.
51. Flavour Lattice Averaging Group. (Y. Aoki *et al.*), *Eur. Phys. J. C* **82**, 869 (2022).
52. D. Leljak, B. Meli and D. van Dyk, *J. High Energy Phys.* **2021**, 36 (2021).
53. A. Bharucha, D. Straub and R. Zwicky, *J. High Energy Phys.* **2016**, 98 (2016).
54. The Muon $g-2$ Collab. (D. Aguillard *et al.*), *Phys. Rev. Lett.* **131**, 161802 (2023).
55. RBC and UKQCD Collab. (T. Blum *et al.*), *Phys. Rev. D* **108**, 054507 (2023).

56. BABAR Collab. (J. Lees *et al.*), *Phys. Rev. D* **86**, 032013 (2012).
57. KLOE-2 Collab. (A. Anastasi *et al.*), *J. High Energy Phys.* **2018**, 173 (2018).
58. V. Druzhinin *et al.*, *Rev. Mod. Phys.* **83**, 1545 (2011).
59. M. Hoferichter *et al.*, *J. High Energy Phys.* **2023**, 08 (2023).
60. Belle II Collab. (I. Adachi *et al.*), *Phys. Rev. D* **110**, 112005 (2024).

QCD Comdensates and α_s from γ-decays: Summary

S. Narison (LUPM Montpellier, FR and iHEPMAD, Univ. Antananarivo, MG)

Light tensor four-quark states from Laplace sum rules[*]

D. Rabetiarivony[†] J. Rakotoarinirina[‡,¶] and H. Rakotonirina[§]
Institute of High-Energy Physics of Madagascar (iHEPMAD),
University of Ankatso, Antananarivo 101, Madagascar
[†]*rd.bidds@gmail.com*
[‡]*razoarinirina109@gmail.com*
[§]*hajarkt0@gmail.com*

Received 25 April 2025
Revised 13 June 2025
Accepted 31 July 2025
Published 11 September 2025

In this work, we use QCD Laplace sum rules (LSR) to estimate the masses and couplings of the $J^P = 2^+ : \rho^+\rho^-, a_2a_0$ and f_2f_0 molecule states. We perform the analysis at lowest order (LO) of the perturbative (PT) series and including the quark and gluon condensates up to dimension $\mathcal{D} \leq 6$. We obtain $M_{\rho^+\rho^-} = 2070 \pm 52$ MeV, $M_{a_2a_0} = 2812 \pm 55$ MeV and $M_{f_2f_0} = 2975 \pm 45$ MeV and for the coupling $f_{\rho^+\rho^-} = 18 \pm 5$ MeV, $f_{a_2a_0} = 83 \pm 8$ MeV and $f_{f_2f_0} = 214 \pm 11$ MeV.

Keywords: QCD spectral sum rules; exotic hadrons; light quark masses.

PACS numbers: 03.65.−w, 04.62.+v

1. Introduction

In previous papers,[1,2] the inverse Laplace transform (LSR)[3–9] of QCD spectral sum rules (QSSR)[a] has been used to extract the masses and couplings of the light scalar four-quark and tensor di-gluonium states. The estimated mass $M_T = 3028(287)$MeV of the 2^{++} di-gluonium state[2] does not favour the interpretation of the $f_2(2010)$, $f_2(2300)$ and $f_2(2340)$ as pure glueball state.

In this work, we pursue our investigation of the tensor state by extending the analysis to the case of the 2^+ $\rho^+\rho^-, a_2a_0$ and f_2f_0 molecule states.

[*]Talk given at the 14th International Conference in High-Energy Physics - HEPMAD24 (21–26th October 2024, Antananarivo–Madagascar).
[¶]Speaker.

[‡]Corresponding author.
[a]For reviews, see Refs. 10–20.

2. The Laplace Sum Rule Approach

For extracting the decay constant and mass of the tensor molecule states, we shall work with the Laplace sum rules

$$\mathcal{L}_0^c(\tau,\mu) \equiv \lim_{Q^2,n\to\infty} \frac{(-Q^2)^n}{(n-1)!}\frac{\partial^n \psi_T}{(\partial Q^2)^n} = \int_{t>}^{t_c} dt\, e^{-t\tau}\frac{1}{\pi}\mathrm{Im}\psi_T(t,\mu), \tag{1}$$

$$\mathcal{R}_{10}^c \equiv \frac{\mathcal{L}_1^c}{\mathcal{L}_0^c} = \frac{\int_{t>}^{t_c} dt\, t\, e^{-t\tau}\mathrm{Im}\psi_T(t,\mu)}{\int_{t>}^{t_c} dt\, e^{-t\tau}\mathrm{Im}\psi_T(t,\mu)}, \tag{2}$$

where τ is the LSR variable, $t_>$ is the hadronic threshold and t_c is the threshold of the "QCD continuum".

The spectral function $\mathrm{Im}\psi_T(t,\mu)$ is evaluated by the calculation of the two-point function:

$$\begin{aligned}\psi_T^{\mu\nu\rho\sigma}(q^2) &\equiv i\int d^4x\, e^{iqx}\langle 0|T\theta^{\mu\nu}(x)(\theta^{\rho\sigma}(0))^\dagger|0\rangle \\ &= \left(P^{\mu\nu\rho\sigma} \equiv \eta^{\mu\rho}\eta^{\nu\sigma} + \eta^{\mu\sigma}\eta^{\nu\rho} - \frac{2}{d-1}\eta^{\mu\nu}\eta^{\rho\sigma}\right)\psi_T(q^2),\end{aligned} \tag{3}$$

where $\theta^{\mu\nu}$ is the interpolating currents describing the tensor molecule states, d is the space–time dimension,

$$\eta^{\mu\nu} \equiv g^{\mu\nu} - q^\mu q^\nu/q^2 \quad \text{and} \quad P_{\mu\nu\rho\sigma}P^{\mu\nu\rho\sigma} = 2(d^2-d-2). \tag{4}$$

To ensure a reliable results, we shall base our analysis on the stability of the masses and couplings versus the change of the external parameters τ and t_c. We shall also implement the rigorous condition:

$$R_{P/C} \equiv \frac{\text{Lowest pole}}{\text{QCD continuum}} = \frac{\int_{t_0}^{t_c} dt\, e^{-t\tau}\rho^{\mathrm{OPE}}(t)}{\int_{t_c}^{\infty} dt\, e^{-t\tau}\rho^{\mathrm{OPE}}(t)} \geq 1, \tag{5}$$

where $\rho^{\mathrm{OPE}}(t) = \frac{1}{\pi}\mathrm{Im}\psi_T(t)$ is the spectral function.

This condition is necessary to better restrict the conservative region of (τ, t_c).

3. $\rho^+\rho^-$, a_2a_0 and f_2f_0 Molecule States

We shall be concerned with the following interpolating currents:

$$\theta_{\mu\nu}^{\rho^+\rho^-} = (\bar u\gamma_\mu d)(u\gamma_\nu \bar d), \tag{6}$$

$$\theta_{\mu\nu}^{a_2a_0} = [\bar u(\gamma_\mu \overleftrightarrow{D}_\nu + \gamma_\nu \overleftrightarrow{D}_\mu)d][u\bar d], \tag{7}$$

$$\theta_{\mu\nu}^{f_2f_0} = \sum_{q=u,d}[\bar q(\gamma_\mu \overleftrightarrow{D}_\nu + \gamma_\nu \overleftrightarrow{D}_\mu)q][\bar u u + \bar d d], \tag{8}$$

where $\overleftrightarrow{D}$ is defined as: $a\overleftrightarrow{D}b = \frac{1}{2}[a(Db) - (Da)b]$.

3.1. *Mass and coupling of* $\rho^+\rho^-$

The expression of the corresponding spectral function reads to LO and up to dimension 6:

$$\rho^{\rm OPE}(t) = \frac{t^4}{7\times 2^{14}\pi^6} - \frac{9\langle\bar qq\rangle(m_d+m_u)t^2}{5\times 2^9\pi^4} - \frac{3\langle g_s^2G^2\rangle t^2}{5\times 2^{13}\pi^6} + \frac{\langle\bar qGq\rangle(m_u+m_d)t}{3\times 2^7\pi^4} + \frac{\langle\bar qq\rangle^2 s}{3\times 2^3\pi^2} - \frac{5\langle g_s^3G^3\rangle s}{3\times 2^{17}\pi^6}. \tag{9}$$

The τ and t_c behavior of the mass and coupling are shown in Fig. 1. The mass presents an inflexion point around $\tau = 0.7 \sim 0.8$ GeV^{-2} which corresponds to a minimum for the coupling. We consider as final and conservative results the mean of the value corresponding to the beginning of τ stability for $[t_c(\text{ GeV}^2), \tau(\text{ GeV}^{-2})] \simeq [10, 0.7]$ and the one where the t_c stability is reached for $[t_c(\text{ GeV}^2), \tau(\text{ GeV}^{-2})] \simeq [18, 0.7]$. One obtains:

$$M_{\rho^+\rho^-} = 2070 \pm 52 \text{ MeV} \quad \text{and} \quad f_{\rho^+\rho^-} = 18 \pm 5 \text{ MeV}. \tag{10}$$

3.2. *Mass and coupling of* a_2a_0

The expression of the spectral function reads to LO and up to dimension 6:

$$\rho^{\rm OPE}(t) = \frac{t^5}{5\times 2^{13}\pi^6} + \frac{133(m_u+m_d)\langle\bar qq\rangle t^3}{45\times 2^{11}\times\pi^4} - \frac{19\langle g_s^2G^2\rangle t^3}{27\times 2^{13}\times\pi^6} - \frac{\langle\bar qGq\rangle s^2}{\pi^4}\left[\frac{11m_d}{5\times 2^9} + \frac{199m_u}{2\times 2^{11}}\right] - \frac{\langle\bar qq\rangle^2 s^2}{45\times 2^5\times\pi^4}. \tag{11}$$

The qualitative behavior of the different curves are very similar to the case of $\rho^+\rho^-$ as shown in Fig. 2. The optimal results are obtained for the set $(t_c, \tau) \simeq (12, 0.7)$ to $(22, 0.8)(\text{ GeV}^2, \text{ GeV}^{-2})$ from which we deduce:

$$M_{a_2a_0} = 2812 \pm 55 \text{ MeV} \quad \text{and} \quad f_{a_2a_0} = 83 \pm 8 \text{ MeV}. \tag{12}$$

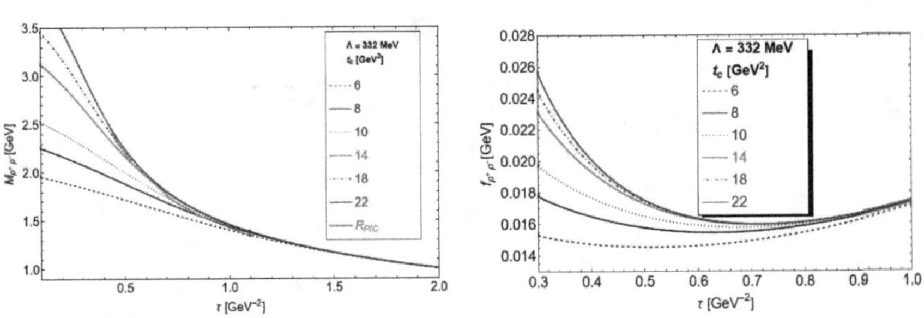

(a) (b)

Fig. 1. (a) $\tau-$behavior of the $\rho^+\rho^-$ mass at LO for different values of t_c. (b) The same caption as (a) but for the coupling $f_{\rho^+\rho^-}$.

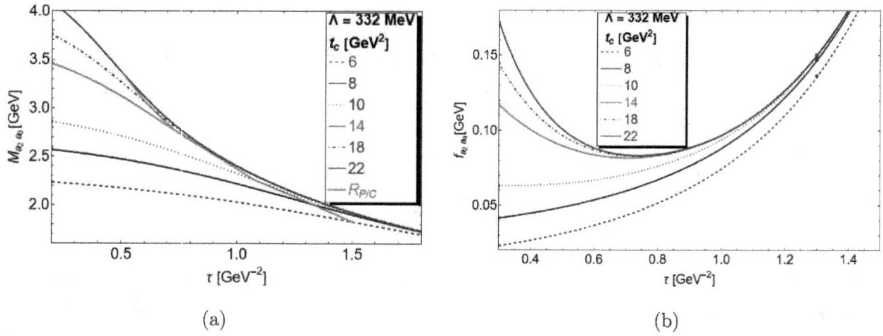

Fig. 2. (a) $\tau-$behavior of the a_2a_0 mass at LO for different values of t_c. (b) The same caption as (a) but for the coupling.

3.3. *Mass and coupling of f_2f_0*

The spectral function reads to LO and up to dimension 6:

$$\rho^{\text{OPE}}(t) = \frac{s^5}{5\times 2^{11}\pi^6} + \frac{37\langle\bar{q}q\rangle(m_u+m_d)t^3}{45\times 2^9\pi^4} + \frac{\langle\bar{q}q\rangle(161m_u+61m_d)t^2}{15\times 2^{12}\pi^4} + \frac{\langle\bar{q}q\rangle^2t^2}{45\times 2^2\pi^4} - \frac{19\langle g_s^2G^2\rangle t^3}{221184\times\pi^6}. \tag{13}$$

The analysis is similar to the previous ones. The optimal results are obtained for the set $(t_c,\tau)\simeq(12,0.7)$ to $(22,0.7)(\text{GeV}^2,\text{GeV}^{-2})$ (Fig. 3) from which we deduce:

$$M_{f_2f_0} = 2975\pm 45\ \text{MeV} \quad \text{and} \quad f_{f_2f_0} = 214\pm 11\ \text{MeV}. \tag{14}$$

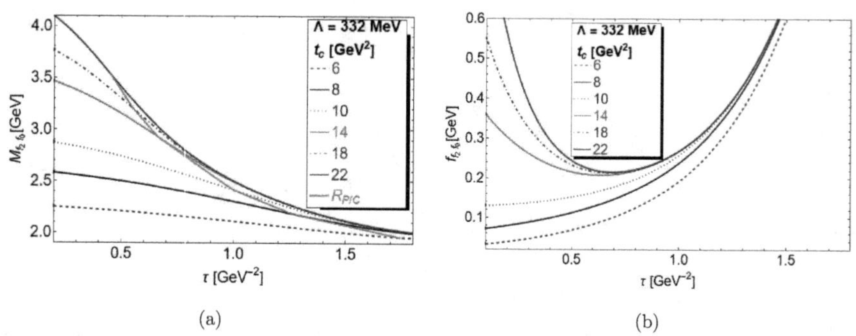

Fig. 3. (a) $\tau-$behavior of the f_2f_0 mass at LO for different values of t_c. (b) The same caption as (a) but for the coupling.

4. Conclusion

We have presented new predictions of QSSR for the masses and couplings of the tensor $\rho^+\rho^-$, a_2a_0 and f_2f_0 molecule states using LSR at LO of PT series and including nonperturbative contribution condensates up to dimension $\mathcal{D} \leq 6$. To better restrict the stability region of (τ, t_c), we have also requested that the Pole contribution is larger than the QCD continuum ($R_{PC} \geq 1$).

In future work, we plan to improve the results of these tensor molecule states by including the Next-to-Leading order (NLO) PT corrections in the analysis.

Acknowledgments

We would like to thank S. Narison for suggesting the subject and partial collaboration.

ORCID

D. Rabetiarivony https://orcid.org/0000-0002-8392-8634
J. Rakotoarinirina https://orcid.org/0009-0005-8534-464X

References

1. R. M. Albuquerque, S. Narison and D. Rabetiarivony, *Nucl. Phys. A* **1039**, 122743 (2023).
2. S. Li, S. Narison, T. Steele and D. Rabetiarivony, *Phys. Lett. B* **849**, 138454 (2024).
3. J. S. Bell and R. A. Bertlmann, *Nucl. Phys. B* **177**, 218 (1981).
4. J. S. Bell and R. A. Bertlmann, *Nucl. Phys. B* **187**, 285 (1981).
5. C. Becchi *et al.*, *Z. Phys. B* **8**, 335 (1981).
6. R. A. Bertlmann, *Acta Phys. Austriaca* **53**, 305 (1981).
7. R. A. Bertlmann and H. Neufeld, *Z. Phys. C* **27**, 437 (1985).
8. S. Narison and E. de Rafael, *Phys. Lett. B* **522**, 266 (1986).
9. M. A. Shifman, A. I. Vainshtein and V. I. Zakharov, *Nucl. Phys. B* **147**, 385 (1979).
10. V. I. Zakharov, *Int. J. Mod. Phys. A* **14**, 4865 (1999).
11. S. Narison, *Nucl. Part. Phys. Proc.* **324–329**, 94 (2023).
12. S. Narison, *QCD as a theory of hadrons, Cambridge Monogr. Part. Phys. Nucl. Phys. Cosmol.* **17**, **1** (2002), arXiv:hep-ph/0205006.
13. S. Narison, *QCD Spectral Sum Rules, World Sci. Lect. Notes Phys.* **26**, 1 (1989).
14. S. Narison, *Phys. Rept.* **84**, 263 (1982); *Acta Phys. Pol. B* **26**, 687 (1995).
15. E. de Rafael, arXiv:hep-ph/9802448.
16. F. J. Yndurain, *The Theory of Quark and Gluon Interactions*, 3rd edn. (Springer, New York, 1999).
17. P. Pascual and R. Tarrach, *QCD: Renormalization for Practitioner* (Springer, New York, 1985).
18. L. J. Reinders, H. Rubinstein and S. Yazaki, *Phys. Rep.* **127**, 1 (1985).
19. B. L. Ioffe, *Prog. Part. Nucl. Phys.* **56**, 232 (2006).
20. H. G. Dosch, *Non-Perturbative Methods*, ed. S. Narison (World Scientific, Singapore, 1985).

2^{++} Di-gluonium from Laplace sum rules at higher order*

Siyuan Li
Department of Physics & Engineering Physics,
University of Saskatchewan, Saskatoon, SK S7N 5E2, Canada
siyuan.li@usask.ca

Stephan Narison
Laboratoire Univers et Particules de Montpellier (LUPM),
CNRS-IN2P3, Case 070, Place Eugène Bataillon, 34095 Montpellier, France
snarison@yahoo.fr

Davidson Rabetiarivony[†,‡]
Institute of High-Energy Physics of Madagascar (iHEPMAD),
University of Antananarivo, Antananarivo 101, Madagascar
rd.bidds@gmail.com

Tom Steele
Department of Physics & Engineering Physics,
University of Saskatchewan, Saskatoon, SK S7N 5E2, Canada
tom.steele@usask.ca

Received 18 February 2025
Accepted 31 July 2025
Published 30 October 2025

We improve the determination of the mass and coupling of the 2^{++} tensor di-gluonium by using relativistic QCD Laplace sum rules (LSRs). In so doing, we evaluate the next-to-leading order (NLO) corrections to the perturbative (PT) and $\langle\alpha_s G^2\rangle$ condensate and the lowest-order (LO) $\langle G^3\rangle$ contributions to the 2^{++} di-gluonium two-point correlator. Within a vacuum saturation estimate ($k_G = 1$) of the dimension-eight gluon condensates, we obtain: $M_T = 3028(287)$ MeV and the renormalization group invariant (RGI) coupling $\hat{f}_T = 224(33)$ MeV. Assuming that the factorization hypothesis can be violated, we study the effect of the violation factor k_G on the results and obtain: $M_T = 3188(337)$ MeV and $\hat{f}_T = 245(32)$ MeV for $k_G = (3 \pm 2)$. Our estimation does not favor the interpretation of the observed $f_2(2010)$, $f_2(2300)$ and $f_2(2340)$ as pure glueball states.

Keywords: QCD spectral sum rules; exotic hadrons; light quark masses.

PACS numbers: 03.65.−w, 04.62.+v

*Talk given at the 14th International Conference in High-Energy Physics — HEPMAD24 (21–26th October 2024, Antananarivo, Madagascar).

[†]Speaker

[‡]Corresponding author.

1. Introduction

In earlier papers,[1–8] we have used the inverse Laplace Transform of QCD spectral sum rules (QSSRs)[9,10] to estimate the masses and decay constants of some light, heavy-light and doubly heavy molecule and tetraquark states.

In this talk, based on the paper in Ref. 11, we improve the lowest-order (LO) estimated mass and coupling of the 2^{++} tensor di-gluonium[10,12–15] by including next-to-leading order (NLO) corrections to the PT and $\langle\alpha_s G^2\rangle$ contributions to the correlation function and studying the effect of the violation of factorization hypothesis for dimension $D = 8$ gluon condensates[16] on the results.

2. The Two-point Correlation Function

We shall work with the two-point function:

$$\psi_T^{\mu\nu\rho\sigma}(q^2) \equiv i\int d^4x\, e^{iqx}\langle 0|T\theta_G^{\mu\nu}(x)(\theta_G^{\rho\sigma}(0))^\dagger|0\rangle$$
$$= \left(P^{\mu\nu\rho\sigma} \equiv \eta^{\mu\rho}\eta^{\nu\sigma} + \eta^{\mu\sigma}\eta^{\nu\rho} - \frac{2}{n-1}\eta^{\mu\nu}\eta^{\rho\sigma}\right)\psi_T(q^2), \tag{1}$$

built from the gluon component of the energy–momentum tensor:

$$\theta_G^{\mu\nu} = \alpha_s\left[-G_\alpha^{\mu,a}G_a^{\nu\alpha} + \frac{1}{4}G_{\alpha\beta}^a G_a^{\alpha\beta}\right], \tag{2}$$

with

$$\eta^{\mu\nu} \equiv g^{\mu\nu} - q^\mu q^\nu/q^2 \quad \text{and} \quad P_{\mu\nu\rho\sigma}P^{\mu\nu\rho\sigma} = 2(n^2-n-2), \tag{3}$$

where $n = 4 + 2\epsilon$ is the space–time dimension used for dimensional regularization and renormalization.

3. QCD Expression of the Two-point Function

3.1. *Lowest-order contribution*

The PT and gluon condensates up to dimension $D = 8$ contributions to the correlation function read

$$\psi_T|_{\text{LO}}(Q^2) = a_s^2\left[-\frac{Q^4}{20}\log\left(\frac{Q^2}{\nu^2}\right) + \frac{\pi^2}{6\alpha_s}\langle\alpha_s G^2\rangle + \frac{5}{3Q^4}\pi^3\alpha_s\langle 2O_1 - O_2\rangle\right], \tag{4}$$

with $a_s = \alpha_s/\pi$ and $O_1 = (f_{abc}G_{\mu\alpha}G_{\nu\alpha})^2, O_2 = (f_{abc}G_{\mu\nu}G_{\alpha\beta})^2$. Assuming vacuum saturation hypothesis ($k_G = 1$), the $D = 8$ gluon condensates can be expressed as

$$\langle 2O_1 - O_2\rangle \simeq -k_G\left(\frac{3}{16}\right)\langle G^2\rangle^2, \tag{5}$$

where k_G is the violation factor. Compared to the expression obtained in Refs. 12 and 13, we found a nonzero $\langle\alpha_s G^2\rangle$ term. However, this term does not affect the analysis as it will disappear when one takes the derivatives of the two-point function.

3.2. *Next-to-leading order contribution*

- PT expression

We shall be concerned with the bare diagrams listed in Table 1 of Ref. 11. We perform the calculation in two ways: diagrammatic renormalization and conventional operator renormalization methods. The diagrammatic renormalization for QCD sum rules correlation functions has been initiated and developed in Ref. 17. The conventional renormalization approach uses the standard Feynman approach[9,10] by considering the renormalization of the gluonic current $\theta_G^{\mu\nu}/\alpha_s$ with the renormalization constant in Ref. 18. The diagrammatic and conventional renormalization approaches lead to the same result. The $\overline{MS}$ renormalized two-point correlation function corrected to NLO for n_f flavors reads

$$\psi_T^{\rm pert}|_{\rm NLO}^R(Q^2) = \psi_T^{\rm pert}|_{\rm LO}\left[1 + a_s\left(\frac{n_f}{6}\log\left(\frac{Q^2}{\nu^2}\right) - \frac{101n_f + 150}{90}\right)\right], \tag{6}$$

where the PT LO term ($\psi_T^{\rm pert}|_{\rm LO}$) can be deduced from Eq. (4). In the case of gluodynamics, we reproduce the results of Ref. 19.

- Dimension-four gluon condensate

We use the renormalization group equation (RGE) to evaluate the leading-log term of $\langle\alpha_s G^2\rangle$ at NLO[11,20,21]:

$$\left\{-\frac{\partial}{\partial t} + \beta(a_s)a_s\frac{\partial}{\partial a_s} - 2\gamma_\psi^\theta\right\}\psi_T^{G^2} = 0, \tag{7}$$

with $t \equiv (1/2)\log(Q^2/\nu^2)$ and γ_ψ^θ the anomalous dimension of the current $\theta_G^{\mu\nu}$:

$$\gamma_\psi^\theta \equiv \left(\gamma_1^\theta = -\frac{11}{2}\right)a_s + \cdots. \tag{8}$$

One obtains

$$\psi_T^{G^2}|_{\rm NLO}(Q^2) = \frac{1}{24\pi}\alpha_s^2\langle\alpha_s G^2\rangle\left(11 + \frac{2n_f}{3}\right)\log\left(\frac{Q^2}{\nu^2}\right). \tag{9}$$

4. 2^{++} Di-gluonium Mass and Coupling at NLO

- The Laplace sum rules

To estimate the mass and decay constant, we shall use the finite energy version of Laplace sum rule (LSR) and their ratios:

$$\mathcal{L}_{0,1}^c(\tau,\nu) \equiv \int_{t_>}^{t_c} dt\ t^{(0,1)}\ e^{-t\tau}\frac{1}{\pi}\mathrm{Im}\ \tilde{\psi}_T(t,\nu);$$
$$\mathcal{R}_{10}^c(\tau) \equiv \frac{\mathcal{L}_1^c}{\mathcal{L}_0^c}. \tag{10}$$

The contribution of the di-gluonium state to the spectral function can be introduced within the Minimal Duality Ansatz (MDA):

$$\frac{1}{\pi}\text{Im } \tilde{\psi}_T(t) = f_T^2 M_T^4 \delta(t - M_T^2) + \theta(t - t_c) \text{ "QCD continuum"}, \quad (11)$$

where f_T is normalized as $f_\pi = 132\,\text{MeV}$. In the MDA

$$\mathcal{R}_{10}^c \simeq M_T^2. \quad (12)$$

Combining Eqs. (4), (6) and (9), one can deduce the LO⊕NLO QCD expression of the correlator up to $D = 8$ gluon condensates normalized to α_s^2: $\tilde{\psi}_T(Q^2)$. The moment $\mathcal{L}_0^c$ is evaluated by taking the third (3rd) derivative of the correlator $\tilde{\psi}_T(Q^2)$ while the $\mathcal{L}_1^c$ moment is obtained by taking the forth (4th) derivative of $Q^2\tilde{\psi}_T(Q^2)$. The expressions of the moments $\mathcal{L}_{0,1}^c$ for $n_f = 3$ flavors to order α_s and up to dimension-eight condensates are

$$\mathcal{L}_0^c = \frac{\tau^{-3}}{10\pi^2}\left\{\left[1 - a_s\left(\frac{53}{15} + \gamma_E\right)\right]\rho_2^c - \frac{65\pi}{12}\langle\alpha_s G^2\rangle\tau^2\rho_0^c \right.$$
$$\left. - \frac{\pi^2}{a_s}\left(\frac{25}{8}\right)k_G\langle\alpha_s G^2\rangle^2\tau^4\right\} \quad (13)$$

and

$$\mathcal{L}_1^c = \frac{3\tau^{-4}}{10\pi^2}\left\{\left[1 - a_s\left(\frac{16}{5} + \gamma_E\right)\right]\rho_3^c - \frac{65\pi}{36}\langle\alpha_s G^2\rangle\tau^2\rho_1^c \right.$$
$$\left. + \frac{\pi^2}{a_s}\left(\frac{25}{24}\right)k_G\langle\alpha_s G^2\rangle^2\tau^4\right\}, \quad (14)$$

with $\gamma_E = 0.5772\ldots$ is the Euler constant and

$$\rho_n^c = 1 - e^{-t_c\tau}\left(1 + (t_c\tau) + \cdots + \frac{(t_c\tau)^n}{n!}\right). \quad (15)$$

- Optimization criteria

We shall use stability criteria on the external variables τ and t_c for extracting the optimal results from the sum rules. The stability regions manifest either as minimum or inflexion point in τ while we shall take t_c in a conservative region from the beginning of τ-stability until the t_c-stability. The subtraction constant ν is eliminated when one takes the different derivatives of the correlator to obtain the inverse Laplace transform and work with the running QCD parameters.

- Mass at NLO

  – *Vacuum saturation hypothesis*

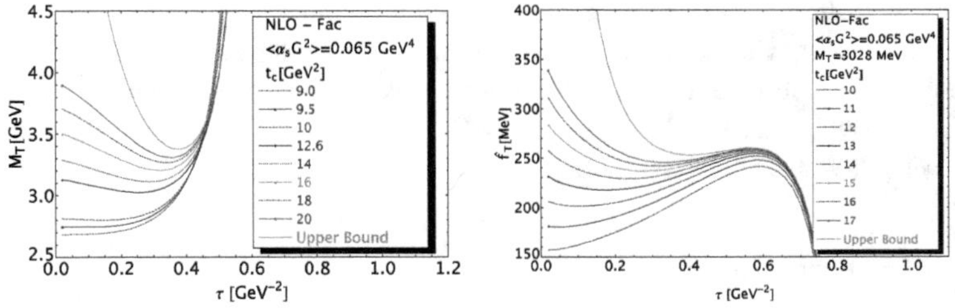

(a) (b)

Fig. 1. (a) τ-behavior of the 2^{++} di-gluonium mass at NLO for different values of t_c where the factorization of $D = 8$ gluon condensates is assumed. (b) The same caption as (a) but for the coupling $\hat{f}_T$.

The behavior of the mass at NLO and assuming factorization ($k_G = 1$) of the dimension-eight gluon condensates is shown in Fig. 1(a). The optimal values of the mass are extracted for the set $(\tau, t_c) = (0.12, 9.5) \sim (0.36, 20)$ $(\text{GeV}^{-2}, \text{GeV}^2)$ where one obtains, respectively: 2746 MeV and 3309 MeV. These values lead to the mean

$$\langle M_T \rangle = 3028(287)\,\text{MeV} \rightsquigarrow t_c \simeq 12.6\,\text{GeV}^2. \tag{16}$$

The optimal upper bound from the positivity of the ratio of moments is

$$M_T \leq [3376(26)_\Lambda(42)_{G^2} = 3376(49)]\,\text{MeV}. \tag{17}$$

– *Violation of vacuum saturation hypothesis*

Assuming that the vacuum saturation hypothesis can be violated ($k_G \neq 1$), we study the effect of the dimension $D = 8$ gluon condensates on the mass determination. Varying the value of the violation factor k_G from 1 to 5, the mass increases slightly to about +319 MeV. Like in the case of the four-quark condensate, we shall use the conservative range $k_G = (3 \pm 2)$ to extract the optimal value of the mass. We obtain

$$\langle M_T \rangle = 3188(337)\,\text{MeV}. \tag{18}$$

• Coupling at NLO

Noting that the current $\theta_G^{\mu\nu}$ acquires an anomalous dimension due to its renormalization[11]; thus we have to introduce the renormalization group invariant (RGI) coupling $\hat{f}_T$ which is related to the running coupling $f_T(\nu)$ as

$$f_T(\nu) = \frac{\hat{f}_T}{\left(\log\left(\frac{\nu}{\Lambda}\right)\right)^{\frac{\gamma_1}{-2\beta_1}}}; \quad \gamma_1 = \frac{n_f}{3}. \tag{19}$$

We shall extract the coupling from the first moment $\mathcal{L}_0^c$.

– *Factorization of $D = 8$ gluon condensates* ($k_G = 1$)

At NLO, as shown in Fig. 1(b), the minimum and inflexion point of the curves are more pronounced. The optimal values are extracted at the τ-minimum which are, respectively, 201 MeV and 246 MeV for the set $(\tau, t_c) = (0.1, 12)$ to $(0.34, 17)$ (GeV^{-2}, GeV^2). We obtain the mean

$$\hat{f}_T|_{\rm NLO} = 224(33){\rm MeV} \rightsquigarrow t_c \simeq 13.5\,{\rm GeV}^2. \tag{20}$$

The optimal upper bound is extracted at the minimum $\tau = 0.4\,{\rm GeV}^{-2}$:

$$\hat{f}_T|_{\rm NLO} \leq 253(1)_\Lambda(2)_{G^2}(32)_{M_T} = 253(32)\,{\rm MeV}. \tag{21}$$

– *Violation of factorization of $D = 8$ gluon condensates*

We use the value of $k_G = (3 \pm 2)$ to perform the analysis. The behavior of the curves is similar to the ones in Fig. 1(b). The optimal values 225 (respectively, 265) MeV are obtained for the set of (τ, t_c): $(0.14, 14)$ (respectively, $(0.36, 22)$) (GeV^{-2}, GeV^2). They lead to the mean

$$\hat{f}_T|_{\rm NLO} = 245(32)\,{\rm MeV} \rightsquigarrow t_c \simeq 7.5\,{\rm GeV}^2. \tag{22}$$

At the inflexion point $\tau = 0.36\,{\rm GeV}^{-2}$, we extract the optimal upper bound:

$$\hat{f}_T|_{\rm NLO} \leq 268(3)_\Lambda(1.5)_{G^2}(30)_{M_T} = 268(32)\,{\rm MeV}. \tag{23}$$

5. Comments and Conclusion

We have presented improved predictions of QSSR of the mass and coupling of 2^{++} tensor di-gluonium state using LSR method. We have evaluated the PT and $D = 4$ gluon condensate NLO corrections and tested the effect of the estimate of the $D = 8$ gluon condensates on the results.

- The PT (respectively, $\langle \alpha_s G^2 \rangle$) NLO corrections increase the mass by about 561 (respectively, 376) MeV from its LO value: 2091 MeV.[11]
- Moving the mass from its LO value to the NLO one: 3028 MeV, the α_s corrections increase the value of the coupling by about 84 MeV. However, if one fixes the mass at its LO value, the coupling is slightly affected by the NLO corrections.
- Our results for NLO $\langle \alpha_s G^2 \rangle$ and LO $\langle G^3 \rangle$ condensate terms do not agree with the ones in Refs. 22 and 23. The discrepancy may come from the different currents used by the authors in Refs. 22 and 23.
- Compared to other approaches, the predicted mass of lattice QCD which is expected to be in the range $(2.27 \sim 2.67)$ GeV[24–26] is slightly lower than ours and the one of instanton liquid model (1525 MeV) is too low. However, our estimated mass goes in line with the errors and the ones of ADS/QCD[27] and some constituent models.[28]
- Confronted to experimental data, our prediction does not favor the interpretation of $f_2(2010)$, $f_2(2300)$ and $f_2(2340)$ states as pure glueball or meson-gluonium mixing[18] candidates.

ORCID

Siyuan Li https://orcid.org/0009-0002-9582-4864
Stephan Narison https://orcid.org/0000-0002-5254-6684
Davidson Rabetiarivony https://orcid.org/0000-0002-8392-8634
Tom Steele https://orcid.org/0000-0003-1716-0783

References

1. R. M. Albuquerque, S. Narison and D. Rabetiarivony, *Nucl. Phys. A* **1039**, 122743 (2023).
2. R. M. Albuquerque, S. Narison and D. Rabetiarivony, *Nucl. Phys. A* **1034**, 122637 (2023).
3. R. M. Albuquerque, S. Narison and D. Rabetiarivony, *Phys. Rev. D* **103**, 074015 (2021).
4. S. Narison and D. Rabetiarivony, *Nucl. Part. Phys. Proc.* **318–323**, 96 (2022).
5. R. M. Albuquerque *et al.*, *Nucl. Phys. A* **1007**, 122113 (2021).
6. R. Albuquerque *et al.*, *Int. J. Mod. Phys. A* **33**, 1850082 (2018).
7. R. Albuquerque *et al.*, *Int. J. Mod. Phys. A* **31**, 1650093 (2016).
8. R. Albuquerque *et al.*, *Int. J. Mod. Phys. A* **31**, 1650196 (2016).
9. S. Narison, *QCD as a Theory of Hadrons*, Cambridge Monographs on Particle Physics, Nuclear Physics and Cosmology, Vol. 17 (Cambridge University Press, Cambridge, 2004), arXiv:hep-ph/0205006.
10. S. Narison, *QCD Spectral Sum Rules*, World Scientific Lecture Notes in Physics, Vol. 26 (World Scientific, Singapore, 1989).
11. S. Li, S. Narison, T. G. Steele and D. Rabetiarivony, *Phys. Lett. B* **849**, 138454 (2024).
12. V. A. Novikov *et al.*, *Nucl. Phys. B* **174**, 378 (1980).
13. V. A. Novikov *et al.*, *Nucl. Phys. B* **191**, 301 (1981).
14. S. Narison, *Z. Phys. C* **22**, 161 (1984).
15. S. Narison, *Nucl. Phys. B* **509**, 312 (1998); *Nucl. Phys. Proc. Suppl.* **64**, 210 (1998).
16. M. A. Shifman, A. I. Vainshtein and V. I. Zakharov, *Nucl. Phys. B* **147**, 385 (1979); ibid. **147**, 448 (1979).
17. T. de Oliveira *et al.*, *Phys. Rev. D* **106**, 114023 (2022).
18. E. Bagan, A. Bramon and S. Narison, *Phys. Lett. B* **196**, 203 (1987).
19. A. A. Pivovarov, *Phys. At. Nucl.* **63**, 1646 (2000); *Yad. Fiz.* **63**, 1734 (2000).
20. S. Narison, *Phys. Rep.* **84**, 263 (1982).
21. D. Asner *et al.*, *Phys. Lett. B* **296**, 171 (1992).
22. H.-X. Chen, W. Chen and S.-L. Zhu, *Phys. Rev. D* **104**, 094050 (2021).
23. H.-X. Chen, W. Chen and S.-L. Zhu, *Nucl. Part. Phys. Proc.* **318–323**, 122 (2022).
24. Y. Chen *et al.*, *Phys. Rev. D* **73**, 014516 (2006).
25. E. Gregory *et al.*, *J. High Energy Phys.* **10**, 170 (2012).
26. A. Athenodorou and M. Teper, *J. High Energy Phys.* **11**, 172 (2020).
27. L. Zhang *et al.*, arXiv:2106.10748 [hep-ph].
28. V. Mathieu, N. Kochelev and V. Vento, *Int. J. Mod. Phys. E* **18**, 1 (2009).

An application of diagrammatic renormalization to 2^{++} tensor di-gluonium

T. de Oliveira,* Siyuan Li† and T. G. Steele‡
Department of Physics and Engineering Physics,
University of Saskatchewan,
Saskatoon, SK, S7N 5E2, Canada
*tdo842@usask.ca
†siyuan.li@usask.ca
‡tom.steele@usask.ca

Received 11 April 2025
Accepted 31 July 2025
Published 8 September 2025

In this paper, we apply the diagrammatic renormalization method to the NLO analysis of the 2^{++} tensor di-gluonium channel within the Quantum chromodynamics (QCD) sum-rules approach. Diagrammatic renormalization eliminates nonlocal divergences directly, avoiding the construction of renormalization factors and complications arising from operator mixing in the conventional renormalization method. The local divergences in QCD correlation functions contribute only to subtraction terms in dispersion relations in QCD sum-rules, making it particularly well-suited for diagrammatic renormalization as the local divergences do not enter sum-rules analysis. We provide a detailed example of renormalizing a representative NLO diagram and perform a comprehensive comparison of all nonzero NLO diagrams for 2^{++} tensor di-gluonium treated with both diagrammatic and conventional operator-mixing methods. The results from both approaches are in agreement, confirming the validity of diagrammatic renormalization. By simplifying the renormalization process, the diagrammatic renormalization method offers a practical alternative for higher-loop analysis of gluonium states and extensions to multi-quark systems.

Keywords: Diagrammatic renormalization; tensor di-gluonium; QCD sum-rules.

PACS numbers: 03.65.−w, 04.62.+v

1. Introduction

As one of the hypothetical exotic hadrons, gluonia have been the subject of ongoing theoretical and experimental searches (see, e.g. Refs. 1–4). In Ref. 5, we studied the $J^{PC} = 2^{++}$ tensor di-gluonium using the QCD Laplace sum-rules approach and improved the theoretical prediction to Next-to-Leading Order (NLO). When evaluating the perturbative contribution to the NLO diagrams for the gluon

†Corresponding author.

current, we encountered challenges related to the renormalization of the Feynman diagram contributions.

In the QCD sum-rules approach, the renormalization of such systems requires the renormalization of correlation functions of composite operators (see Refs. 6 and 7 for overview on QCD sum-rules). In the conventional renormalization method, where counter-terms are added to the Lagrangian density and additional renormalization factors are needed for composite operators, the renormalization of those composite operators results in the operator-mixing, and becomes progressively more complicated as the mass dimension of the operator increases, limiting the calculation power in QCD sum-rules applications (see Refs. 8–10 for discussions on QCD renormalization).

The diagrammatic renormalization method,[11–14] however, has been shown to avoid the explicit mixing of composite operators and increase the efficiency of the renormalization for scalar and vector mesonic correlation functions for light and heavy quarks, light scalar quark meson and scalar glueball mixed correlation function, for heavy-non-strange and heavy-strange diquark correlation functions, and for non-strange 0^{+-} tetraquark systems.[15–17] In this paper, we introduce the 2^{++} tensor di-gluonium correlation function in Sec. 2. The diagrammatic renormalization method is presented and applied to the perturbative contribution of the 2^{++} tensor di-gluonium correlator in Sec. 3. The correlation function renormalized via the diagrammatic renormalization method is compared to the conventional method of renormalization in Sec. 4.

2. 2^{++} Tensor Di-Gluonium Correlation Function

Using Laplace sum-rules method, the authors of Ref. 5 studied the NLO correction to 2^{++} Tensor Di-gluonium including both perturbative and nonperturbative contributions of the two-point correlation function[3]:

$$\psi^T_{\mu\nu\rho\sigma} \equiv i \int d^4x e^{iqx} \left\langle 0 \middle| T J^g_{\mu\nu}(x) J^g_{\rho\sigma}(0)^\dagger \middle| 0 \right\rangle = P_{\mu\nu\rho\sigma} \psi_T(q^2), \tag{1}$$

where $P_{\mu\nu\rho\sigma}$ is the spin-2 projection operator at $D = 4 + 2\epsilon$ space–time dimension (see Refs. 1, 3 and 18):

$$P_{\mu\nu\rho\sigma} = \eta_{\mu\rho}\eta_{\nu\sigma} + \eta_{\mu\sigma}\eta_{\nu\rho} - \frac{2}{D-1}\eta_{\mu\nu}\eta_{\rho\sigma}, \text{ with } \eta_{\mu\nu} \equiv g_{\mu\nu} - \frac{q_\mu q_\nu}{q^2}, \tag{2}$$

whose normalization factor is

$$P_{\mu\nu\rho\sigma}P^{\mu\nu\rho\sigma} = 2(D^2 - D - 1), \tag{3}$$

and $J^g_{\mu\nu}(x)$ are the gluonic currents that are built from the gluon fields[3]:

$$J^g_{\mu\nu}(x) = -G^{\alpha,a}_{\mu}(x)G_{\nu\alpha,a}(x) + \frac{1}{4}g_{\mu\nu}G^a_{\alpha\beta}(x)G^{\alpha\beta}_a(x)\,. \tag{4}$$

The Leading-Order (LO) contribution to Eq. (1) is a Feynman diagram with a one-loop gluonic self-energy topology. The NLO perturbative contributions to the correlation function (1) are listed in the second column of Tables 1 and 2. Applying QCD Laplace sum-rules methods, all local divergencies from (1) are removed via a Borel transform and all nonlocal divergences are removed via the diagrammatic renormalization method.

3. Diagrammatic Renormalization

In the diagrammatic renormalization method (see, e.g. Refs. 10 and 15 for a review and applications of the method) each Feynman diagram G from the perturbative expansion of Eq. (1) is individually renormalized by subtracting counter-term diagrams $C(G)$ from the bare diagram $U(G)$. To construct each $C(G)$, isolate a sub-diagram from G (i.e. a subset of lines that contains at least one loop), extract the divergent term γ from the sub-diagram (the sub-divergence), and then replace the sub-diagram by γ in G. The renormalized diagram $R(G)$ is obtained by subtracting the counter-term diagrams associated with each sub-diagram, $C_\gamma(G)$, from the bare diagram $U(G)$

$$R(G) = U(G) - \sum_\gamma C_\gamma(G). \tag{5}$$

After each diagram G from the correlation function is renormalized through Eq. (5), the sum of all $R(G)$ is the renormalized correlation function with all nonlocal divergences removed from Eq. (1) (local divergencies are removed in QCD sum-rule methods) and with coupling and mass parameters interpreted as their renormalized versions. An example of the 2^{++} tensor di-gluonium NLO perturbative evaluation of the bare diagram for $\psi_T(q^2)$ along with the diagrammatic renormalization procedure is provided in Sec. 3.1.

3.1. *Example: Gluon self-energy with gluonic loop*

The nonzero contribution to Eq. (1) for the gluon self-energy topology (diagram (ii) in Table 1) is shown in Fig. 1 and is given by (the applied Feynman rules can be

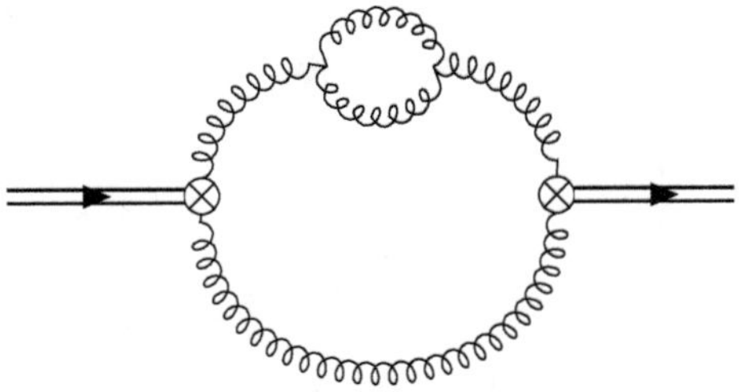

Fig. 1. NLO gluon-self energy with gluonic loop nonzero contribution to the correlation function (1), where the curly lines stands for gluon fields, the ⊗ denotes the composite operator insertion, and the double line represents the external momentum q.

referred to in Refs. 8, 19 and 20):

$$\Pi^{(ii)}_{\text{bare}}(Q^2 \equiv -q^2) = \frac{19\alpha_s Q^4}{1600\pi^3} L\left(10L - \frac{2642}{57} + \frac{10}{\epsilon}\right), L = \log\left(\frac{Q^2}{\nu^2}\right), \tag{6}$$

where the result is in Feynman gauge, ν is the MS scheme renormalization scale, and local divergences (polynomials in Q^2) are omitted from Eq. (6).

The diagram in Fig. 1 has two sub-diagrams, represented in the third column of Table 1 and in Figs. 2(b) and 2(d). Figures 2(b) and 2(d) are the loops indicated by the dashed boxes in Figs. 2(a) and 2(c), respectively. Once the sub-diagrams are isolated, the counter-term diagrams (see the fourth column in Table 1) are generated by replacing the dashed boxes in Figs. 2(a) and 2(c) by the corresponding divergent term from Figs. 2(b) and 2(d), resulting in Figs. 3(a) and 3(b), respectively.

The counter-term in Fig. 3(a) is a massless tadpole. The loop integral of such topology is zero in dimensional regularization.[8] Even in the case of a massive tadpole, the external momentum q does not enter the loop integral for Fig. 3(a). Hence, the tadpole diagram corresponds to a dispersion-relation subtraction constant in Eq. (1)

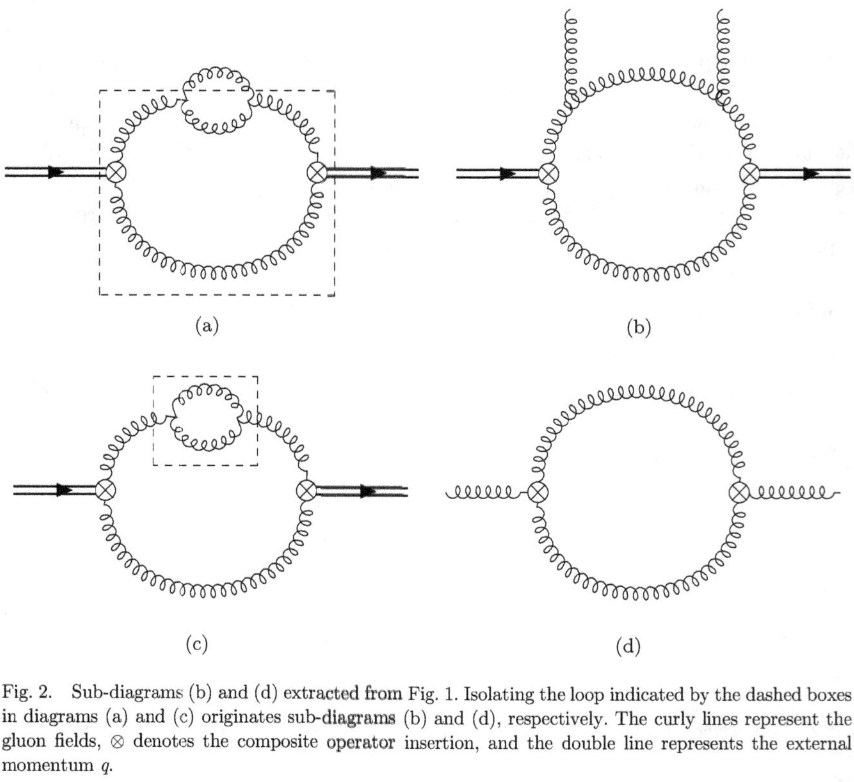

Fig. 2. Sub-diagrams (b) and (d) extracted from Fig. 1. Isolating the loop indicated by the dashed boxes in diagrams (a) and (c) originates sub-diagrams (b) and (d), respectively. The curly lines represent the gluon fields, $\otimes$ denotes the composite operator insertion, and the double line represents the external momentum q.

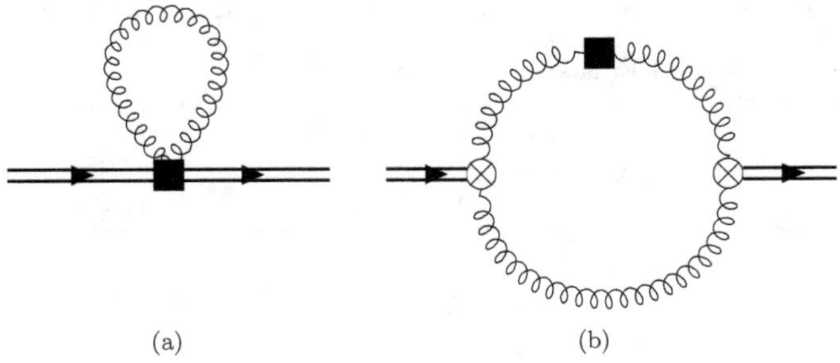

Fig. 3. Counter-term diagrams generated by the sub-diagrams of Fig. 2(b) and 2(d). The curly lines represent the gluon fields, $\otimes$ denotes the composite operator insertion, the square ■ represents the sub-divergence insertion, and the double line represents the external momentum q.

and does not have a physical contribution to QCD sum-rules, representing a local divergence. This type of counter-term diagram also shows up for the bare diagrams (iii), (iv), (v) and (vi) in Tables 1 and 2. For the counter-term diagram in Fig. 3(a), the sub-divergence γ insertion, represented by the square ■, from sub-diagram (d) in Fig. 2 is given by

$$\gamma = i\delta^{ab}\frac{g^2}{64\pi^2\epsilon}\left(22q^\mu p^\nu - 19p^2 g^{\mu\nu}\right), \tag{7}$$

where the Latin letters represent color indices. Note that the $\frac{1}{\epsilon}$ divergence subtraction are performed in MS scheme. So, the renormalized result will need conversion to $\overline{\text{MS}}$ scheme using $\nu^2 \to \exp(\gamma_E)\nu^2/4\pi$. The resulting contribution $\Pi^{(\text{ii})}_{\text{ct}}$ from Fig. 3(b) is then

$$\Pi^{(\text{ii})}_{\text{ct}}(Q^2 \equiv -q^2) = \frac{19\alpha_s Q^4}{1600\pi^3}L\left(5L - 9 + \frac{10}{\epsilon}\right),\ L = \log\left(\frac{Q^2}{\nu^2}\right). \tag{8}$$

Note that the nonlocal divergent L/ϵ terms from Eqs. (6) and (8) are identical. According to the diagrammatic renormalization method, the renormalized diagram can be found by subtracting the counter-term diagram contribution given in Eq. (8) from the bare diagram contribution given in Eq. (6). Therefore, for the diagram in Fig. 1, the renormalized result is

$$\begin{aligned}\Pi^{(\text{ii})}_{\text{renorm}}(Q^2) &= \Pi^{(\text{ii})}_{\text{bare}}(Q^2) - \Pi^{(\text{ii})}_{\text{ct}}(Q^2)\\ &= \frac{\alpha_s Q^4}{4800\pi^3}L(285L - 2129),\ L = \log\left(\frac{Q^2}{\nu^2}\right).\end{aligned} \tag{9}$$

3.2. *Renormalized correlation function*

To obtain the complete NLO 2^{++} tensor di-gluonium renormalized correlation function, one must sum the contributions from all the individual NLO renormalized

Table 1. NLO (nonzero) bare diagrams, sub-diagrams and counter-term diagrams for the 2^{++} tensor di-gluonium in the diagrammatic renormalization procedure. Curly lines represent the gluon fields, dotted lines represent the ghost field, solid lines represent the quark field, ⊗ is the composite operator insertion, ■ is the sub-divergence insertion, and the double line represents the external momentum q. Diagram symmetry-related cases are taken into account and incorporated into the final results.

Label	NLO Bare diagrams	Sub-Diagrams	Counter-term Diagrams
(i)	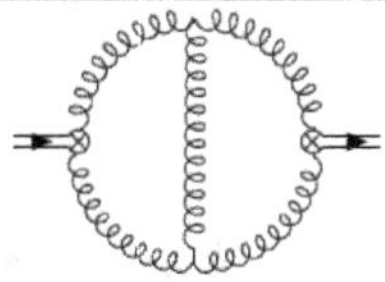	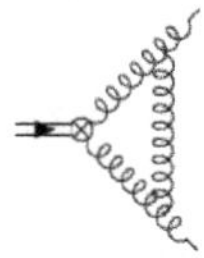	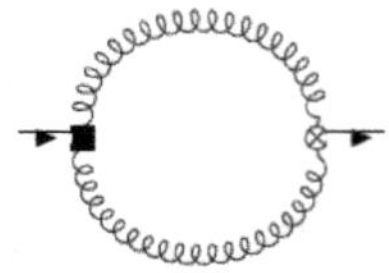
(ii)	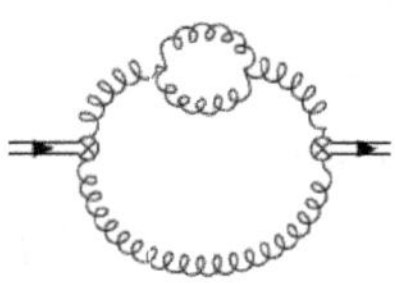 +	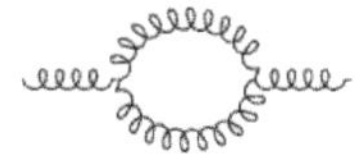	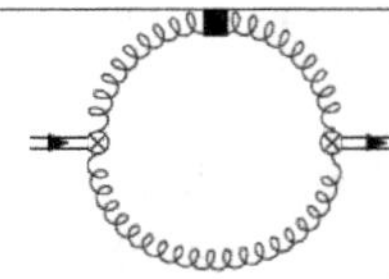
	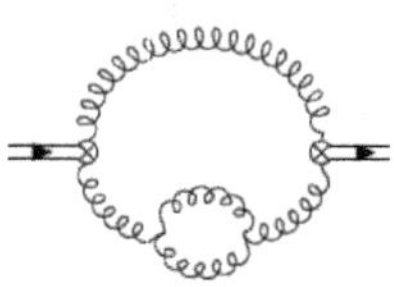	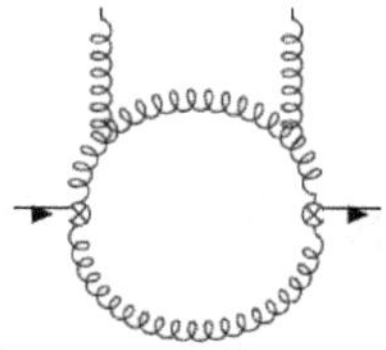	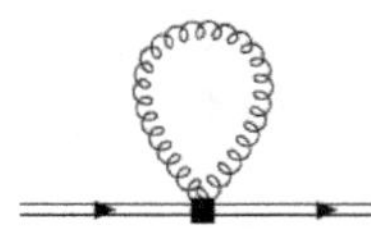
(iii)	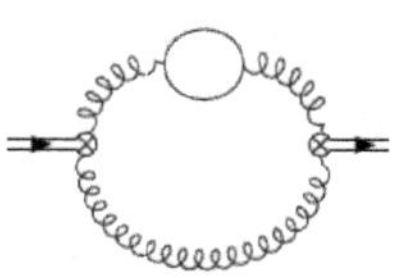 +	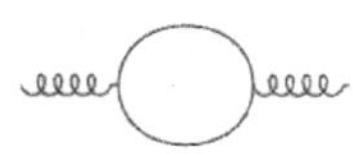	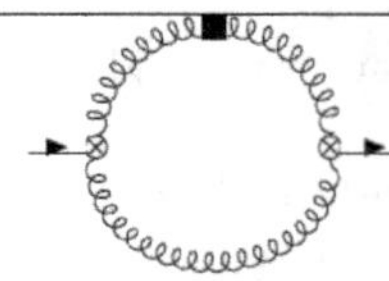
	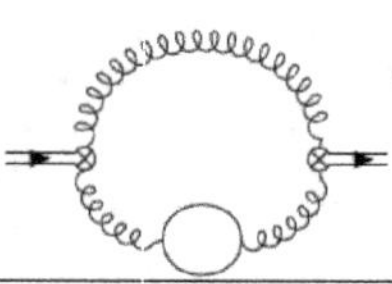	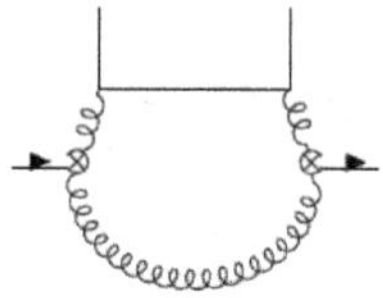	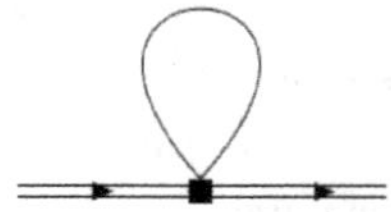
(iv)	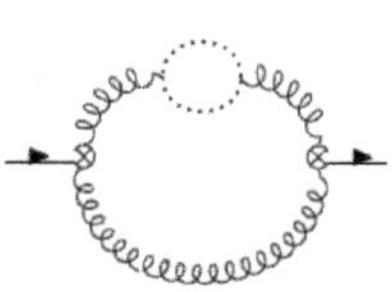 +		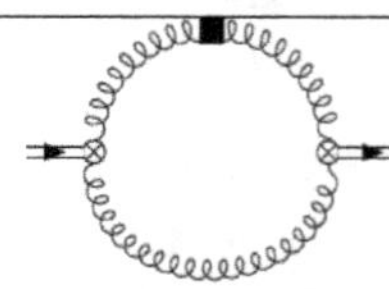
	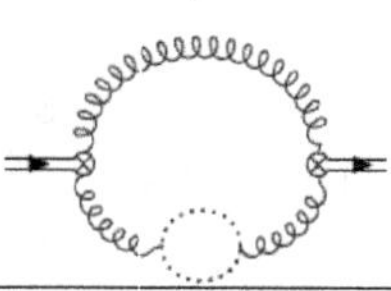	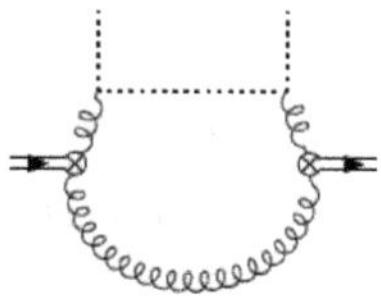	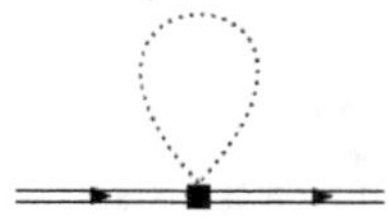

Table 2. NLO (nonzero) bare diagrams, sub-diagrams and counter-term diagrams for the 2^{++} tensor di-gluonium in the diagrammatic renormalization procedure. Curly lines represent the gluon fields, dotted lines represent the ghost field, solid lines represent the quark field, ⊗ is the composite operator insertion, ■ is the sub-divergence insertion, and the double line represents the external momentum q. Diagram symmetry-related cases are taken into account and incorporated into the final results.

Label	NLO Bare Diagrams	Sub-Diagrams	Counter-term Diagrams
(v)	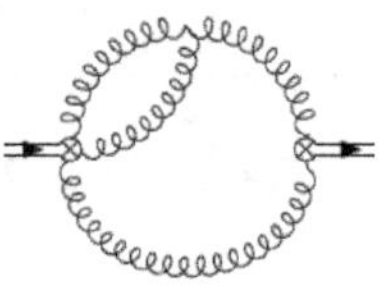 + 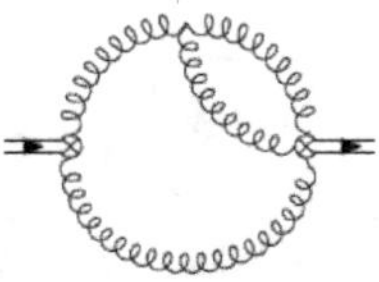	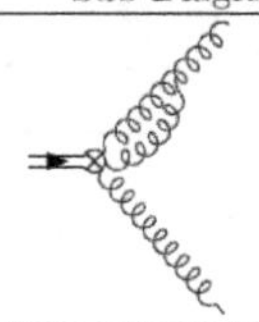	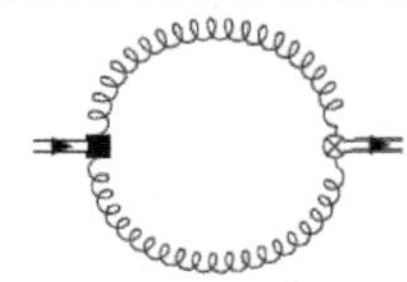
		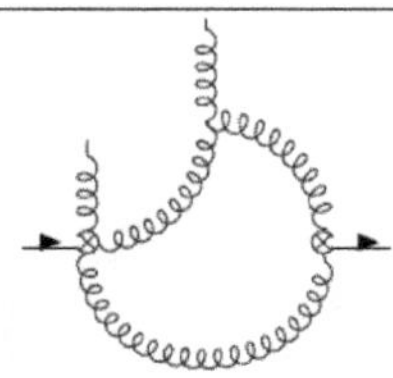	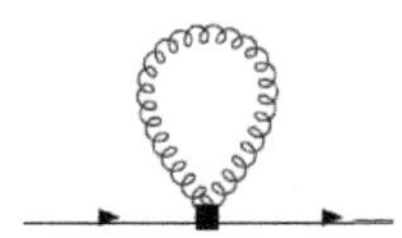
(vi)	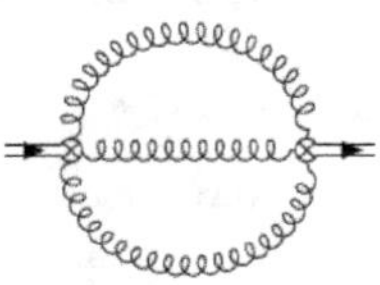	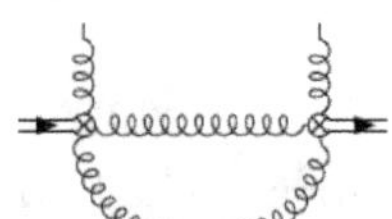	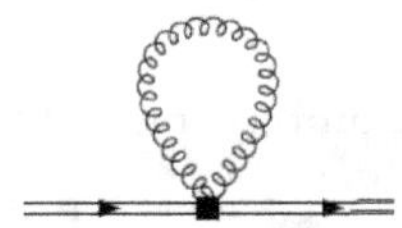
(vii)	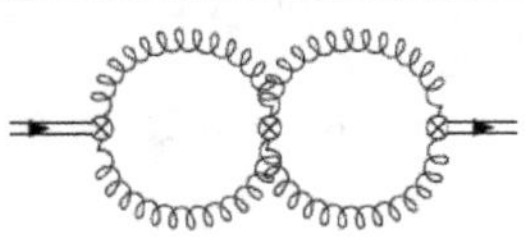	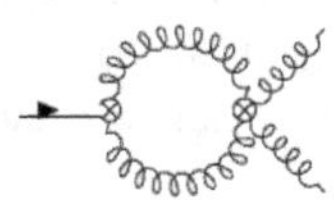	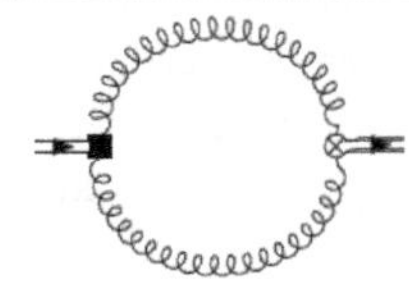

diagrams shown in Tables 1 and 2:

$$\Pi^{NLO}_{\text{renorm}}(Q^2) = \sum_{m=a}^{g} \Pi^{(m)}_{\text{renorm}} = \frac{\alpha_s}{1800\pi^3} Q^4 L\big[(101 - 15L)n_f + 150\big],\ L = \log\left(\frac{Q^2}{\nu^2}\right), \tag{10}$$

for n_f flavors.

The contributions of each diagram — bare, counter-term, and renormalized — are summarized in Table 3, while the corresponding diagram representations are presented in Tables 1 and 2. Diagram symmetry-related cases are taken into account and incorporated into the final results in Table 3.

Table 3. NLO (perturbative) bare, subtraction and counterterm diagrams for the 2^{++} tensor di-gluonium contribution to $\psi_T(q^2)$ in Eq. (1), evaluated using both the conventional and diagrammatic renormalization methods. A and B are defined by $\psi_T(Q^2) = (\frac{\alpha_s}{\pi})^3(\frac{Q^4}{16})L(AL + B)$ for each diagram's contribution. n_f stands for number of flavors. Diagram (viii) is used exclusively in the conventional renormalization approach, as detailed in Sec. 4. All symmetry-related diagram cases have been evaluated and incorporated into the final results.

	Bare		Counter-term		Renormalized	
Label	A	B	A	B	A	B
(i)	$-\frac{7}{30}$	$\frac{3107}{1800} - \frac{7}{30\epsilon}$	$-\frac{7}{60}$	$\frac{521}{225} - \frac{7}{30\epsilon}$	$-\frac{7}{60}$	$-\frac{1151}{1800}$
(ii)	$\frac{19}{10}$	$-\frac{1321}{150} + \frac{19}{10\epsilon}$	$\frac{19}{20}$	$-\frac{171}{100} + \frac{19}{10\epsilon}$	$\frac{19}{20}$	$-\frac{2129}{300}$
(iii)	$-\frac{4}{15}n_f$	$\frac{256}{225}n_f - \frac{4}{15\epsilon}n_f$	$-\frac{2}{15}n_f$	$\frac{6}{25}n_f - \frac{4}{15\epsilon}n_f$	$-\frac{2}{15}n_f$	$\frac{202}{225}n_f$
(iv)	$\frac{1}{10}$	$-\frac{79}{150} + \frac{1}{10\epsilon}$	$\frac{1}{20}$	$-\frac{9}{100} + \frac{1}{10\epsilon}$	$\frac{1}{20}$	$-\frac{131}{300}$
(v)	0	−1	0	0	0	−1
(vi)	$-\frac{31}{10}$	$\frac{2887}{200} - \frac{31}{10\epsilon}$	$-\frac{31}{20}$	$\frac{128}{75} - \frac{31}{10\epsilon}$	$-\frac{31}{20}$	$\frac{7637}{600}$
(vii)	$\frac{4}{3}$	$-\frac{40}{9} + \frac{4}{3\epsilon}$	$\frac{2}{3}$	$-\frac{20}{9} + \frac{4}{3\epsilon}$	$\frac{2}{3}$	$-\frac{20}{9}$
(viii)	$\frac{2}{15}n_f$	$-\frac{6}{25}n_f + \frac{4}{15\epsilon}n_f$	N/A	N/A	N/A	N/A
Total	$-\frac{2}{15}n_f$	$\frac{202}{225}n_f + \frac{4}{3}$			$-\frac{2}{15}n_f$	$\frac{202}{225}n_f + \frac{4}{3}$

4. Comparison to Conventional Renormalization Method

The diagrammatic renormalization method is particularly convenient for handling the renormalization of individual diagrams. However, when determining the overall NLO contribution to the 2^{++} tensor di-gluonium, the conventional approach involves incorporating renormalization-induced diagrams.

As discussed in Ref. 1, the renormalized meson-gluonium currents take the following operator-mixing forms:

$$\begin{aligned} J^{\mu\nu}_{q,R} &= Z_{11}J^{\mu\nu}_{q,B} + Z_{12}J^{\mu\nu}_{g,B}, \\ J^{\mu\nu}_{g,R} &= Z_{21}J^{\mu\nu}_{q,B} + Z_{22}J^{\mu\nu}_{g,B}, \end{aligned} \tag{11}$$

where the renormalization constants Z_{ij} are in $\overline{\text{MS}}$ scheme, and $J^{\mu\nu}_{q,B}$ is the bare quark current. To $\mathcal{O}(\alpha_s/\pi)$, the only renormalization constant needed is Z_{22} that connects the bare gluon current $J^{\mu\nu}_{g,B}$ to renormalized gluon current $J^{\mu\nu}_{g,R}$:

$$Z_{22} = 1 + R_{22}, \text{ with } R_{22} = -\frac{\alpha_s}{6\pi\epsilon}n_f. \tag{12}$$

R_{22} will construct a new diagram term, called the renormalization-induced diagram, to renormalized the divergences from the carried bare diagrams. R_{22} will be embedded in a term which is proportional to the current-gluon vertex, represented as the shaded circle in Fig. 4. Due to the symmetry of the tensor gluonium LO diagram, there are two renormalization-induced diagrams, represented in Figs. 4(a) and 4(b), both yielding identical contributions. The sum of both contributions gives the following:

$$\Pi^{(viii)}_{\text{bare}}(Q^2 \equiv -q^2) = \frac{\alpha_s Q^4}{600\pi^3}n_f L\left(5L + \frac{10}{\epsilon} - 9\right),\ L = \log\left(\frac{Q^2}{\nu^2}\right). \tag{13}$$

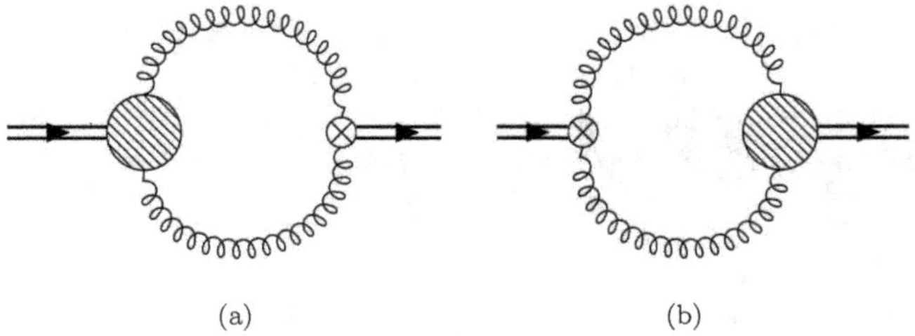

Fig. 4. Renormalization induced-diagrams in the conventional operator-mixing approach. Curly lines represent the gluon fields, ⊗ denotes the composite operator insertion, the shaded circle represents R_{22} contribution, and the double line represents the external momentum q.

The renormalized 2^{++} tensor di-gluonium is obtained by summing the NLO bare diagrams and the renormalization-induced diagrams determined by R_{22}. The result from the conventional renormalization method matches the one derived using the diagrammatic renormalization method as given in Eq. (10), and shown in the last row of Table 3. This agreement serves as a strong validation of the diagrammatic renormalization method when applied to the 2^{++} tensor di-gluonium.

5. Conclusion

Diagrammatic renormalization offers an alternative to conventional renormalization techniques for QCD correlation functions. This approach effectively eliminates nonlocal divergences, thereby circumventing the intricate construction of renormalization factors for composite operators — particularly those involving operator mixing. It is especially well-suited for QCD sum-rule analyses, where local divergences manifest solely as subtraction terms in the dispersion relation and thus do not contribute to the sum-rule itself. As a result, the renormalization process is significantly simplified, requiring attention only to the logarithmic (nonlocal) divergences.

In this paper, we provided a detailed example within the QCD sum-rule framework by applying diagrammatic renormalization to a NLO diagram of the 2^{++} tensor di-gluonium in Sec. 3. A comprehensive comparison of all relevant NLO diagrams renormalized via both diagrammatic and conventional methods is presented in Sec. 4 and summarized in Table 3 (see also Refs. 5 and 20). We find that diagrammatic renormalization yields results in full agreement with the conventional approach. By bypassing the complexities of operator mixing, diagrammatic renormalization emerges as a viable and efficient method for studying higher-loop corrections in gluonium and, more broadly, multi-quark systems within QCD sum-rules.[15]

Acknowledgment

TGS is grateful for research funding from the Natural Sciences and Engineering Research Council of Canada (NSERC).

ORCID

T. de Oliveira https://orcid.org/0000-0003-3976-1272
Siyuan Li https://orcid.org/0009-0002-9582-4864
T. G. Steele https://orcid.org/0000-0003-1716-0783

References

1. E. Bagan, A. Bramon and S. Narison, *Phys. Lett. B* **196**, 203 (1987), doi:10.1016/0370-2693(87)90604-6.
2. Y. Chen, A. Alexandru, S. J. Dong, T. Draper, I. Horváth, F. X. Lee, K. F. Liu, N. Mathur, C. Morningstar, M. Peardon, S. Tamhankar, B. L. Young and J. B. Zhang, *Phys. Rev. D* **73**, 014516 (2006), doi:10.1103/PhysRevD.73.014516.
3. S. Narison, *Nucl. Phys. B* **509**, 312 (1998); *Nucl. Phys. B Proc. Suppl.* **64**, 210 (1998).
4. BESIII Collaboration (M. Ablikim *et al.*), *Phys. Rev. Lett.* **132**, 181901 (2024), doi:10.1103/PhysRevLett.132.181901.
5. S. Li, S. Narison, T. Steele and D. Rabetiarivony, *Phys. Lett. B* **849**, 138454 (2024), doi:10.1016/j.physletb.2024.138454.
6. M. Shifman, A. Vainshtein and V. Zakharov, *Nucl. Phys. B* **147**, 385 (1979), doi:10.1016/0550-3213(79)90022-1.
7. M. Shifman, A. Vainshtein and V. Zakharov, *Nucl. Phys. B* **147**, 448 (1979), doi:10.1016/0550-3213(79)90023-3.
8. P. Pascual and R. Tarrach, *QCD: Renormalization for the Practitioner*, *Lecture Notes in Physics*, Vol. 194 (Springer-Verlag, Berlin, Heidelberg, 1984).
9. S. Narison, *QCD as a Theory of Hadrons: From Partons to Confinement* (Oxford University Press, 2005).
10. J. C. Collins, *Renormalization: An Introduction to Renormalization, the Renormalization Group and the Operator-Product Expansion*, *Cambridge Monographs on Mathematical Physics* (Cambridge University Press, 1984).
11. N. N. Bogoliubov and D. V. Shirkov, *Introduction to the Theory of Quantized Fields* (Wiley, New York, 1980).
12. K. Hepp, *Commun. Math. Phys.* **2**, 301 (1966), doi:10.1007/BF01773358.
13. W. Zimmermann, *Commun. Math. Phys.* **15**, 208 (1969), doi:10.1007/BF01645676.
14. J. C. Collins, *Nucl. Phys. B* **92**, 477 (1975), doi:10.1016/S0550-3213(75)80010-1.
15. T. de Oliveira, D. Harnett, A. Palameta and T. G. Steele, *Phys. Rev. D* **106**, 114023 (2022), doi:10.1103/PhysRevD.106.114023.
16. T. de Oliveira, D. Harnett, R. Kleiv, A. Palameta and T. G. Steele, *Phys. Rev. D* **108**, 054036 (2023), doi:10.1103/PhysRevD.108.054036.
17. K. Ray, D. Harnett and T. G. Steele, *Phys. Rev. D* **108**, 034001 (2023), doi:10.1103/PhysRevD.108.034001.
18. J. Govaerts, L. J. Reinders, P. Francken, X. Gonze and J. Weyers, *Nucl. Phys. B* **284**, 674 (1987), doi:10.1016/0550-3213(87)90056-3.
19. T. P. Cheng and L. F. Li, *Gauge Theory of Elementary Particle Physics* (Oxford University Press, Oxford, UK, 1984).
20. S. Li, *Precision Calculations in Theoretical Particle Physics and the Search for New Physics*, Ph.D. thesis, University of Saskatchewan (2025).

XYZ from QCD Laplace sum rules

A. Rabemananjara [*] and D. Rabetiarivony [†]
Institute of High-Energy Physics of Madagascar,
University of Antananarivo,
Antananarivo 101, Madagascar
[*] *achrisrab@gmail.com*
[†] *rd.bidds@gmail.com*

Received 18 April 2025
Revised 5 August 2025
Accepted 6 August 2025
Published 11 September 2025

In this paper, we present a review of our recent works on heavy-light molecule and tetraquark states. New compact integrated expressions of the spectral functions at lowest order (LO) of perturbative (PT) QCD and up to dimension $d \leq 6-8$ condensates of the Operator Product Expansion (OPE) are used. Including higher-order (HO) PT corrections which we have estimated by assuming the factorization of the molecule/four-quark spectral functions, we evaluate masses and couplings with the Laplace sum rules (LSR). Our optimal results based on stability criteria are summarized in Tables 4 to 6 and compared with experimental candidates and some LO QCD spectral sum rules (QSSR) results. We conclude that the masses of the observed XZ states are compatible with (almost) pure $J^{PC} = 1^{+\pm}, 0^{++}$ molecule or/and four-quark states. The SU(3) corrections on the meson masses are tiny: $\leq 10\%$ (respectively $\leq 3\%$) for the c (respectively b)-quark channel but can be large for the couplings ($\leq 20\%$).

Keywords: QCD spectral sum rules; perturbative and nonperturbative QCD; exotic hadrons; masses and decay constants.

PACS numbers: 03.65.−w, 04.62.+v

1. Introduction

This paper summarizes the results in our original works.[1,2]

A large amount of exotic hadrons which differ from the "standard" $\bar{c}c$ chamonium and $\bar{b}b$ bottomium radial excitation states have been recently discovered.[3–10,a] Most of them have been assigned as four-quark and/or molecule states.[11] In Refs. 1 and 2, we have systematically extracted the masses and couplings of some possible heavy-light

[*] Corresponding author.
[a] More details and complete references are in Sec. 1 of Refs. 1 and 2.

($\bar{Q}\bar{q}Qq$) configurations using the inverse Laplace transform (LSR) version of QSSR.[b] We included to the heavy-light exotic correlators the next-to-next-to-leading-order (N2LO) factorized PT corrections in the chiral limit and adding to these the SU(3) next-to-leading order (NLO) factorized PT corrections. We have emphasized the importance of the pertubative (PT) corrections for giving a meaning on the input heavy quark mass value which plays an important role in the analysis.

2. The Laplace Sum Rule (LSR)

We shall work with the Finite Energy version of the QCD Inverse Laplace sum rules and their ratios:

$$\mathcal{L}_n^c(\tau,\mu) = \int_{(2M+2m)^2}^{t_c} dt\; t^n e^{-t\tau}\frac{1}{\pi}\mathrm{Im}\Pi_{\mathcal{M},\mathcal{T}}(t,\mu); \quad M_{\mathcal{H}}^2 = \mathcal{R}_n^c = \frac{\mathcal{L}_{n+1}^c}{\mathcal{L}_n^c}\bigg|_{n=0}, \tag{1}$$

where M and m are the on-shell/pole heavy and running light quark masses, τ is the LSR variable, $n = 0; 1$ is the degree of moments, t_c is the threshold of the QCD continuum which parametrizes, from the discontinuity of the Feynman diagrams, the spectral function evaluated by the calculation of the correlator defined as

$$\Pi_{\mathcal{M},\mathcal{T}}(t,\mu)(q^2) = i\int d^4x\; e^{iqx}\langle 0|T[\mathcal{O}_{\mathcal{M},\mathcal{T}}^J(x)(\mathcal{O}_{\mathcal{M},\mathcal{T}}^J)^\dagger(0)]|0\rangle, \tag{2}$$

where $O_{\mathcal{M},\mathcal{T}}^J(x)$ are the interpolating currents for the tetraquarks $\mathcal{T}$ and molecules $\mathcal{M}$ states. The superscript J refers to the spin of the particles. We have extracted the lowest ground state mass by using the minimal duality ansatz (MDA):

$$\frac{1}{\pi}\mathrm{Im}\Pi_{\mathcal{M},\mathcal{T}} \simeq f_{\mathcal{M},\mathcal{T}}^2 M_{\mathcal{M},\mathcal{T}}^8 \delta(t - M_{\mathcal{M},\mathcal{T}}^2) + \Theta(t - t_c). \tag{3}$$

Double ratios of sum rules (DRSR)[16] are also useful for extracting the SU3 breaking effects on couplings and masses:

$$r_{\mathcal{H}'/\mathcal{H}} = \sqrt{\mathcal{R}_{\mathcal{H}'}^c/\mathcal{R}_{\mathcal{H}}^c} = \frac{M_{\mathcal{H}'}}{M_{\mathcal{H}}}. \tag{4}$$

Variables τ, t_c and μ are, in principle, free external parameters. We shall use stability criteria with respect to these free 3 parameters to extract the lowest ground state mass and coupling (more detailed discussions can be seen in Refs. 1 and 2 and references therein).

3. Higher PT Corrections to the Spectral Function

As an example, we will use the following scalar $\bar{D}D$ (0^{++}) molecule current:

$$\mathcal{O}_{\bar{D}D} = (\bar{q}\gamma_5 c)(\bar{c}\gamma_5 q). \tag{5}$$

[b] For reviews, see e.g. Refs. 12–15 and references therein.

We extract higher PT corrections by considering that the molecule/tetraquark two-point spectral function is the convolution of the two ones built from two quark bilinear currents (factorization).[17,18] In this way, for the current in (5), we have the following:

$$\frac{1}{\pi}\mathrm{Im}\psi^{(s)}_{mol}(t) = \theta(t-4M_c^2)\left(\frac{1}{4\pi}\right)^2 t^2 \times \int_{M_c^2}^{(\sqrt{t}-M_c)^2} \mathrm{d}t_1 \int_{M_c^2}^{(\sqrt{t}-\sqrt{t_1})^2} \mathrm{d}t_2\, \lambda^{1/2}\, \mathcal{K}^{\bar{\mathrm{D}}\mathrm{D}}, \quad (6)$$

where M_c is the on-shell/pole perturbative charm quark mass.

$$\mathcal{K}^{\bar{D}D} \equiv \left(\frac{t_1}{t}+\frac{t_2}{t}-1\right)^2 \times \frac{1}{\pi}\mathrm{Im}\psi^{(p)}(t_1)\frac{1}{\pi}\mathrm{Im}\psi^{(p)}(t_2), \quad (7)$$

with the phase space factor:

$$\lambda = \left(1-\frac{(\sqrt{t_1}-\sqrt{t_2})^2}{t}\right)\left(1-\frac{(\sqrt{t_1}+\sqrt{t_2})^2}{t}\right). \quad (8)$$

4. QCD Input Parameters

The QCD parameters[c] which shall appear in the analysis are α_s the strange and heavy quark masses (the light quark masses have been neglected). We shall also consider nonperturbative condensates. Their values are given in Table 1.

The Renormalization Group Invariant parameters are defined as[13,14]

$$\begin{gathered}\bar{m}_s(\tau) = \hat{m}_s(-\beta_1 a_s)^{-2/\beta_1}, \quad \langle\bar{q}q\rangle(\tau) = -\hat{\mu}_q^3(-\beta_1 a_s)^{2/\beta_1}, \\ \langle\bar{q}Gq\rangle(\tau) = -M_0^2\hat{\mu}_q^3(-\beta_1 a_s)^{1/3\beta_1},\end{gathered} \quad (9)$$

where $\hat{\mu}_q$ is the spontaneous RGI light quark condensate.[36] The running mass m is related to the on-shell (pole) mass M from the NLO relation:[37–41]

$$M(\mu) = \bar{m}(\mu)\left[1+\frac{4}{3}a_s(\mu)+\log\left(\frac{\mu}{M}\right)^2 a_s(\mu)+\mathcal{O}(a_s^2)\right]. \quad (10)$$

Table 1. QCD input parameters estimated from QSSR (Moments sum rules, LSR and ratios of sum rules).

Parameters	Values	Refs.
$\alpha_s(M_Z)$	0.1181(16)(3)	19, 20
$\hat{m}_s$	(0.114 ± 0.006) GeV	14, 21, 22
$\overline{m}_c(m_c)$	1266(6) MeV	19, 23
$\overline{m}_b(m_b)$	4202(8) MeV	23
$\hat{\mu}_q$	(253 ± 6) MeV	14, 21, 22
M_0^2	(0.8 ± 0.2) GeV^2	14, 24–29
$\langle\alpha_s G^2\rangle$	$(6.35 \pm 0.35) \times 10^{-2}$ GeV^4	19
$\langle g^3 G^3\rangle$	(8.2 ± 2.0) GeV$^2 \times \langle\alpha_s G^2\rangle$	30–32
$\rho\alpha_s\langle\bar{q}q\rangle^2$	$(5.8 \pm 0.9) \times 10^{-4}$ GeV^6	25, 26, 33–35

[c]Note that the values of $\overline{m}_b(m_b) = 4177(11)$ MeV and $\langle\alpha_s G^2\rangle = (7 \pm 3) \times 10^{-2}$ GeV^4 in the original papers[1,2] are obsolete values.

5. Application of MDA and Stability Criteria

We use the scalar $\bar{D}D$ (0^{++}) molecule current in Eq. (5). The μ-dependence of the results for given $t_c = 32\,\text{GeV}^2$ and $\tau \sim 0.35\,\text{GeV}^{-2}$ is shown in Fig. 1. Using the fact that the final result must be independent of the arbitrary parameter μ (plateau/ inflection point for the coupling and minimum for the mass), we consider as an optimal result the one at $\mu = 4.5\,\text{GeV}$.

We study the τ-behavior of the coupling from the moment $\mathcal{L}_0^c$ and mass from the ratio of moments $\mathcal{R}_0^c$ for different values of t_c at NLO (respectively N2LO) as shown in Fig. 2 (respectively Fig. 3). We consider as final and conservative results the mean of the value corresponding to the beginning of τ-stability for $[t_c(\text{GeV}^2), \tau(\text{GeV}^{-2})] \simeq [22,\ 0.25]$ and the one where t_c-stability starts to be reached for $[t_c(\text{GeV}^2), \tau(\text{GeV}^{-2})] \simeq [32,\ 0.35]$. In these stability regions, the requirement that the pole contribution is larger than the one of the continuum is automatically satisfied.

Using the value of $t_c = 32\,\text{GeV}^2$, we study in Fig. 4 the convergence of the PT series for a given value of $\mu = 4.5$ GeV. We observe that from NLO to N2LO the mass decreases by about only 1 per mil indicating the good convergence of the PT-series.

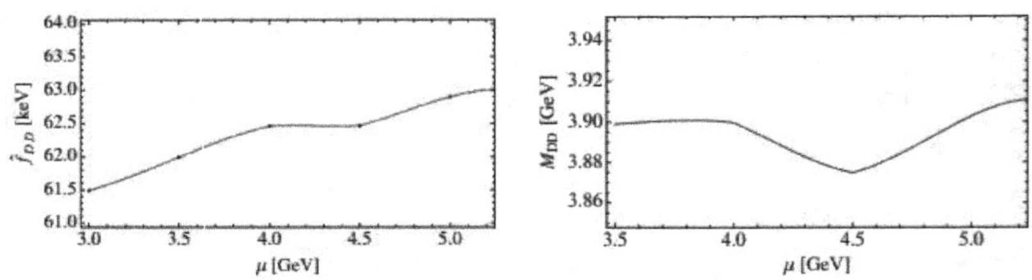

(a) (b)

Fig. 1. (a) Renormalization group invariant coupling $\hat{f}_{\bar{D}D}$ at NLO as function of μ, for the corresponding τ-stability region, for $t_c = 32\ \text{GeV}^2$ and for the QCD parameters in Table 1; (b) the same as (a) but for the mass $M_{\bar{D}D}$.

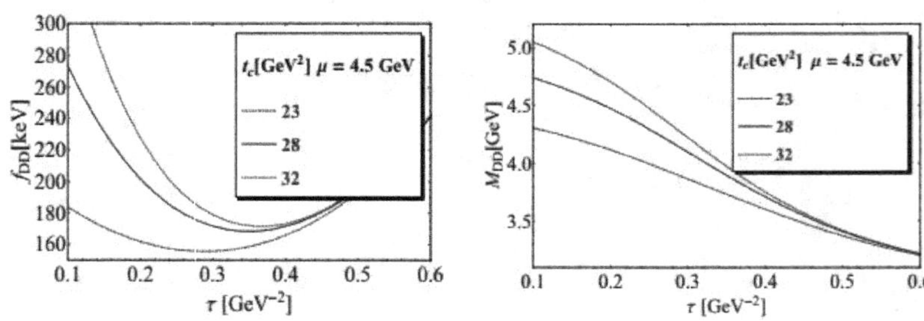

(a) (b)

Fig. 2. (a) $f_{\bar{D}D}$ at NLO as function of τ for different values of t_c, for $\mu = 4.5$ GeV and for the QCD parameters in Table 1; (b) the same as (a) but for the mass $M_{\bar{D}D}$.

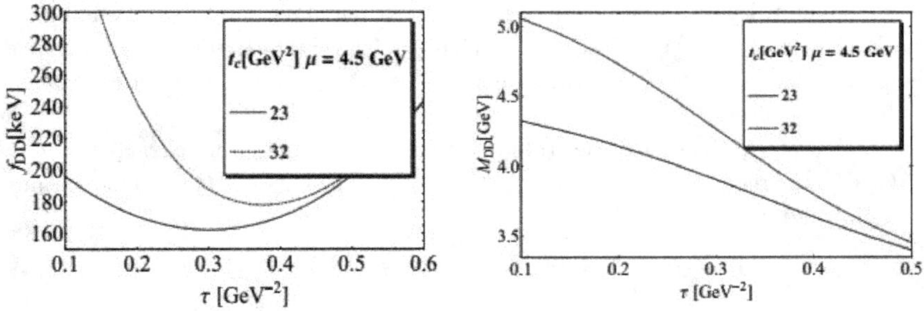

(a) (b)

Fig. 3. (a) $f_{\bar{D}D}$ at N2LO as function of τ for different values of t_c, for $\mu = 4.5$ GeV and for the QCD parameters in Table 1; (b) same caption as (a) but for the mass $M_{\bar{D}D}$.

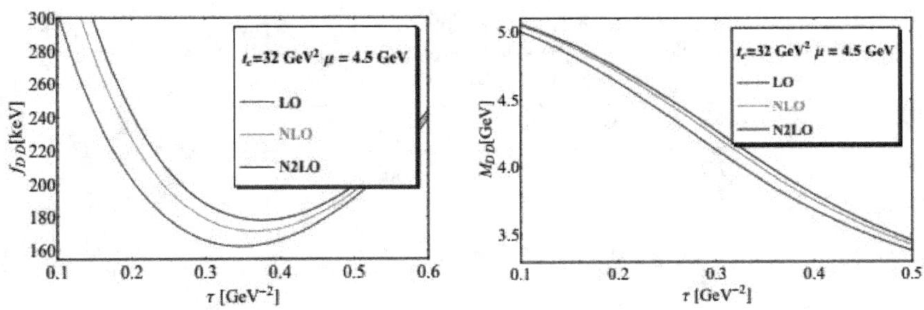

(a) (b)

Fig. 4. (a) $f_{\bar{D}D}$ as function of τ for a given value of $t_c = 32$ GeV^2, for $\mu = 4.5$ GeV, for different truncation of the PT series and for the QCD parameters in Table 1; (b) the same as (a) but for the mass $M_{\bar{D}D}$.

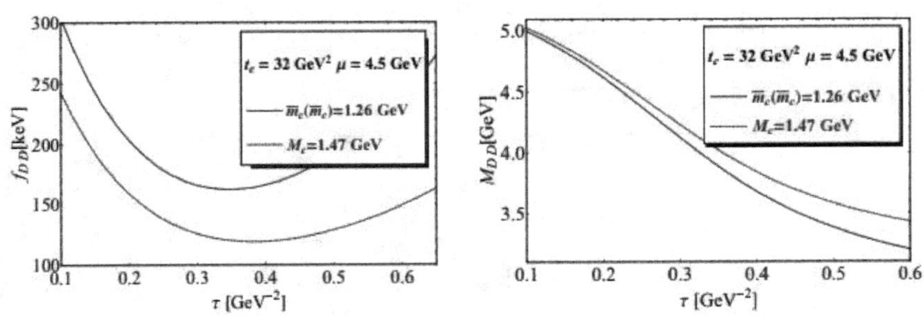

(a) (b)

Fig. 5. (a) $f_{\bar{D}D}$ at LO as function of τ for $t_c = 32$ GeV^2, for $\mu = 4.5$ GeV, for values of the running $\bar{m}_c(\bar{m}_c) = 1.26$ GeV and pole mass $M_c = 1.47$ GeV. We use the QCD parameters in Table 1; (b) the same as (a) but for the mass $M_{\bar{D}D}$.

We show in Fig. 5 the effect of the definitions (running and pole) of the heavy quark mass used in the analysis at LO which is relatively important. The difference should be added as errors in the LO analysis. This source of errors is never considered in the current literature.

Finally, for the $\bar{D}D$ molecule state, the final results have been extracted at $\mu = 4.5\,\text{GeV}$ in the region of $\tau = 0.25 \sim 0.35\,\text{GeV}^{-2}$ and $t_c = 22 \sim 32\,\text{GeV}^2$. One obtains the following:

$$M_{\bar{D}D} = 3898(36)\ \text{MeV}, \quad f_{\bar{D}D} = 170(15)\ \text{keV}. \tag{11}$$

6. *XYZ*-like Spectra at N2LO in the Chiral Limit

We pursue and complete the previous study by reconsidering the existing estimates of the masses of the *XYZ*-like states. In so doing, we use the interpolating currents in Tables 2 and 3. The complete results are compiled in Table 4.

Table 2. Interpolating currents with a definite C-parity describing the molecule-like states. $Q \equiv c$ (respectively b) for the $\bar{D}D$ (respectively $\bar{B}B$)-like molecules, $q \equiv u$, d.

States	J^{PC}	Currents
$\bar{D}D$	0^{++}	$(\bar{q}\gamma_5 Q)(\bar{Q}\gamma_5 q)$
$\bar{D}^*D^*$		$(\bar{q}\gamma_\mu Q)(\bar{Q}\gamma^\mu q)$
$\bar{D}_0^*D_0^*$		$(\bar{q}Q)(\bar{Q}q)$
$\bar{D}^*D$	1^{++}	$\frac{i}{\sqrt{2}}[(\bar{Q}\gamma_\mu q)(\bar{q}\gamma_5 Q) - (\bar{q}\gamma_\mu Q)(\bar{Q}\gamma_5 q)]$
$\bar{D}_0^*D_1$		$\frac{i}{\sqrt{2}}[(\bar{q}Q)(\bar{Q}\gamma_\mu\gamma_5 q) + (\bar{Q}q)(\bar{q}\gamma_\mu\gamma_5 Q)]$
$\bar{D}_0^*D$	$0^{-\pm}$	$\frac{i}{\sqrt{2}}[(\bar{q}Q)(\bar{Q}\gamma_5 q) \pm (\bar{Q}q)(\bar{q}\gamma_5 Q)]$
$\bar{D}^*D_1$		$\frac{i}{\sqrt{2}}[(\bar{Q}\gamma_\mu q)(\bar{q}\gamma^\mu\gamma_5 Q) \mp (\bar{Q}\gamma_\mu\gamma_5 q)(\bar{q}\gamma_\mu Q)]$
$\bar{D}_0^*D^*$	$1^{-\pm}$	$\frac{i}{\sqrt{2}}[(\bar{q}Q)(\bar{Q}\gamma_\mu q) \mp (\bar{Q}q)(\bar{q}\gamma_\mu Q)]$
$\bar{D}D_1$		$\frac{i}{\sqrt{2}}[(\bar{Q}\gamma_\mu\gamma_5 q)(\bar{q}\gamma_5 Q) \pm (\bar{q}\gamma_\mu\gamma_5 Q)(\bar{Q}\gamma_5 q)]$

Table 3. Interpolating currents with a definite C-parity describing the $\bar{D}_sD_s$ molecule-like states and their beauty analogue.

States	J^{PC}	Currents
$\bar{D}_sD_s$	0^{++}	$(\bar{s}\gamma_5 Q)(\bar{Q}\gamma_5 s)$
$\bar{D}_s^*D_s^*$		$(\bar{s}\gamma_\mu Q)(\bar{Q}\gamma^\mu s)$
$\bar{D}_{s0}^*D_{s0}^*$		$(\bar{s}Q)(\bar{Q}s)$
$\bar{D}_{s1}^*D_{s1}^*$		$(\bar{s}\gamma_\mu\gamma_5 Q)(\bar{Q}\gamma^\mu\gamma_5 s)$
$\bar{D}_s^*D_s$	1^{++}	$\frac{i}{\sqrt{2}}[(\bar{Q}\gamma_\mu s)(\bar{s}\gamma_5 Q) - (\bar{s}\gamma_\mu Q)(\bar{Q}\gamma_5 s)]$
$\bar{D}_{s0}^*D_{s1}$		$\frac{i}{\sqrt{2}}[(\bar{s}Q)(\bar{Q}\gamma_\mu\gamma_5 s) + (\bar{Q}s)(\bar{s}\gamma_\mu\gamma_5 Q)]$
$\bar{D}_{s0}^*D_s$	$0^{-\pm}$	$\frac{i}{\sqrt{2}}[(\bar{s}Q)(\bar{Q}\gamma_5 s) \pm (\bar{Q}s)(\bar{s}\gamma_5 Q)]$
$\bar{D}_s^*D_{s1}$		$\frac{i}{\sqrt{2}}[(\bar{Q}\gamma_\mu s)(\bar{s}\gamma^\mu\gamma_5 Q) \mp (\bar{Q}\gamma_\mu\gamma_5 s)(\bar{s}\gamma_\mu Q)]$
$\bar{D}_{s0}^*D_s^*$	$1^{-\pm}$	$\frac{i}{\sqrt{2}}[(\bar{s}Q)(\bar{Q}\gamma_\mu s) \mp (\bar{Q}s)(\bar{s}\gamma_\mu Q)]$
$\bar{D}_sD_{s1}$		$\frac{i}{\sqrt{2}}[(\bar{Q}\gamma_\mu\gamma_5 s)(\bar{s}\gamma_5 Q) \pm (\bar{s}\gamma_\mu\gamma_5 Q)(\bar{Q}\gamma_5 s)]$

Table 4. *DD*-like molecules and four-quarks masses and couplings from LSR at N2LO.

States	M(MeV)	f(keV)	States	M(MeV)	f(keV)
$\mathbf{0^{++}}$			$\mathbf{0^{++}}$		
$\bar{D}D$	3898(36)	170(15)	BB	10595(58)	17(4)
$\bar{D}^*D^*$	3903(179)	302(47)	$\bar{B}^*B^*$	10647(184)	32(5)
$\bar{D}_0^*D_0^*$	4398(54)	136(22)	$\bar{B}_0^*B_0^*$	10648(113)	14(4)
$\mathbf{1^{+\pm}}$			$\mathbf{1^{+\pm}}$		
$\bar{D}^*D$	3903(62)	161(17)	$\bar{B}^*B$	10646(150)	17(5)
$\bar{D}_0^*D_1$	4401(164)	137(16)	$\bar{B}_0^*B_1$	10692(132)	14(2)
$\mathbf{0^{-\pm}}$			$\mathbf{0^{-\pm}}$		
$\bar{D}_0^*D$	5690(140)	257(19)	$\bar{B}_0^*B$	12562(260)	67(10)
$\bar{D}^*D_1$	5797(141)	564(38)	$\bar{B}^*B_1$	12627(225)	136(19)
$\mathbf{1^{--}}$			$\mathbf{0^{--}}$		
$\bar{D}_0^*D^*$	5748(101)	261(17)	$\bar{B}_0^*B^*$	12592(266)	66(11)
$\bar{D}D_1$	5544(162)	231(21)	BB_1	12573(257)	65(12)
$\mathbf{1^{-+}}$			$\mathbf{1^{-+}}$		
$\bar{D}_0^*D^*$	5828(132)	249(22)	$\bar{B}_0^*B^*$	12617(220)	64(8)
$\bar{D}D_1$	5748(179)	241(25)	BB_1	12630(236)	65(9)
$S_c(0^+)$	3898(54)	191(20)	$S_b(0^+)$	10654(109)	19(4)
$A_c(1^+)$	3888(130)	184(30)	$A_b(1^+)$	10680(172)	19(3)
$\pi_c(0^-)$	5750(127)	310(13)	$\pi_b(0^-)$	12695(254)	94(11)
$V_c(1^-)$	5793(122)	296(19)	$V_b(1^-)$	12610(242)	75(9)

7. *XYZ*-SU3 Breakings at Higher Orders

The LHCb collaboration confirmed the existence of the 1^{++} $X(4147)$ and $X(4273)$ found earlier by the CDF,[7] CMS[8] and D0[9] collaborations. In the same time, the LHCb collaboration has reported the existence of the 0^{++} states in the $J/\varphi\phi$ invariant mass and the BELLE[10] collaboration discovered the $Y(4351)$ with a width of 13 MeV.

Motivated by these discoveries, we pursue our investigation by including analogous N2LO PT corrections in the chiral limit and adding to these the SU3 next-to-leading order (NLO) PT corrections to the heavy-light correlators. We use DRSR for extracting the SU3 breaking effects on couplings and masses. Our results are compiled in Tables 5 and 6.

Table 5. $\bar{D}D$-like molecules and four-quarks couplings, masses and their corresponding SU3 ratios from LSR at N2LO of PT.

States	f_M^{sd}	f_{M_s}(keV)	r_M^{sd}	M_{M_s}(MeV)
$\mathbf{0^{++}}$				
$\bar{D}_sD_s$	0.98(4)	167(18)	1.070(4)	4169(48)
$\bar{D}_s^*D_s^*$	0.95(3)	284(34)	1.075(3)	4196(200)
$\bar{D}_{s0}^*D_{s0}^*$	0.89(6)	102(14)	1.058(68)	4225(132)
$\bar{D}_{s1}^*D_{s1}$	0.930(34)	229(31)	1.090(7)	4124(61)
$\bar{D}_0^*D_0^*$		114(18)		3954(224)
$\bar{D}_1D_1$		274(37)		3784(56)

Table 5. (*Continued*)

States	f_M^{sd}	f_{M_s}(keV)	r_M^{sd}	M_{M_s}(MeV)
$1^{+\pm}$				
$\bar{D}_s^*D_s$	0.97(3)	156(17)	1.073(4)	4188(67)
$\bar{D}_{s0}^*D_{s1}$	0.82(1)	110(18)	1.100(24)	4275(206)
$\bar{D}_0^*D_1$		112(17)		3854(182)
$0^{-\pm}$				
$\bar{D}_{s0}^*D_s$	0.90(4)	232(25)	0.946(40)	5385(214)
$\bar{D}_s^*D_{s1}$	0.90(4)	508(38)	0.972(34)	5632(192)
1^{--}				
$\bar{D}_{s0}^*D_s^*$	0.86(4)	216(11)	0.956(32)	5571(180)
$\bar{D}_sD_{s1}$	0.93(3)	213(13)	0.951(31)	5272(120)
1^{-+}				
$\bar{D}_{s0}^*D_s^*$	0.92(5)	231(18)	0.948(32)	5528(179)
$\bar{D}_sD_{s1}$	0.88(3)	212(14)	0.955(34)	5487(152)
$S_{sc}(0^+)$	0.98(4)	187(19)	1.086(11)	4233(61)
$A_{sc}(1^+)$	0.87(4)	160(17)	1.082(4)	4209(112)
$\pi_{sc}(0^-)$	0.86(7)	267(30)	0.96(3)*	5524(176)
$V_{sc}(1^-)$	0.87(10)	258(33)	0.96(4)*	5539(234)

The $*$ indicates that the value does not come from a direct determination. ($f_M^{sd} \equiv f_{M_s}/f_M$, $r_M^{sd} \equiv M_{M_s}/M_M$).

Table 6. $\bar{B}B$-like molecules and four-quarks couplings, masses and their corresponding SU3 ratios from LSR at N2LO of PT.

States	f_M^{sb}	f_{M_s}(keV)	r_M^{sb}	M_{M_s}(MeV)
0^{++}				
B_sB_s	1.15(4)	20(2)	1.029(4)	10906(74)
$\bar{B}_s^*B_s^*$	1.12(3)	36(6)	1.029(5)	10956(134)
$\bar{B}_{s0}^*B_{s0}^*$	1.07(5)	17(4)	1.034(11)	11014(224)
$\bar{B}_{s1}^*B_{s1}$	1.214(74)	29(6)	1.035(2)	10882(169)
$\bar{B}_1B_1$		28.6(4)		10514(149)
$1^{+\pm}$				
$\bar{B}_s^*B_s$	1.18(4)	20(2)	1.030(4)	10972(195)
$\bar{B}_{s0}^*B_{s1}$	0.79(4)	11(3)	1.031(14)	11021(204)
$0^{-\pm}$				
$\bar{B}_{s0}^*B_s$	1.02(3)	68(4)	1.00(3)*	12509(213)
$\bar{B}_s^*B_{s1}$	0.95(4)	118(13)	1.00(3)*	12573(292)
0^{--}				
$\bar{B}_{s0}^*B_s^*$	0.90(3)	59(5)	0.99(3)*	12512(263)
$\bar{B}_sB_{s1}$	0.77(3)	50(3)	0.99(3)*	12426(233)
1^{-+}				
$\bar{B}_{s0}^*B_s^*$	0.92(3)	59(6)	0.99(3)*	12479(257)
$\bar{B}_sB_{s1}$	0.85(3)	55(6)	0.98(3)*	12350(242)
$S_{sb}(0^+)$	0.83(3)	26(6)	1.048(4)	11133(149)
$A_{sb}(1^+)$	0.98(3)	26(5)	1.046(6)	11172(172)
$\pi_{sb}(0^-)$	0.76(4)	71(13)	0.975(2)*	12374(209)
$V_{sb}(1^-)$	0.90(6)	68(9)	0.984(30)*	12411(266)

8. Comments and Conclusion

Our results show that radiative corrections are relatively smaller for the masses than for the couplings in some channel, which can explain the agreement of our results for the masses with the LO ones given in the literature.

- *Chiral limit*
  - For the nonstrange states found from their decays into $J/\psi\pi^+\pi^-$, one can conclude that the $X_c(3872)$ and $Z_c(3900)$ can be well described with an almost pure $\bar{D}^*D$ molecule or/and four-quark $[\bar{c}\bar{q}cq]$ states ($q \equiv u, d$). The $Z_c(4200, 4430)$ might be a $\bar{D}_0^*D_1$ molecule state.
  - The $J^{PC} = 0^{++}$ $\bar{D}D$, $\bar{D}^*D^*$ molecule and four-quark nonstrange states are almost degenerated with the 1^{++} ones (mass around 3900 MeV).
  - The predicted mass of the $\bar{D}_0^*D_0^*$ molecule is higher but is still below the threshold. The $1^{-\pm}$ molecules and four-quark states in the range of $(5646 \sim 5961)$ MeV are too high compared with the observed $Y_c(4140)$ to $Y_c(4660)$ states.
  - The $0^{-\pm}$ molecules populate the region $(5656 \sim 6020)$ MeV. Like in the case of vector states, these pseudoscalar states are well above the physical threshold. Therefore, these molecule states should be broad and are difficult to separate from backgrounds.
  - For the axial-vector 1^{++} states, the value of $\sqrt{t_c} \simeq (6 \sim 7)$ GeV where the optimal values of the masses have been extracted is approximately the mass of the first radial excitation. One can conclude that the higher masses experimental states cannot be such radial excitation.
  - For the b-channel, the $Y_b(9898, 10260, 10870)$ are too low compared to the predicted unmixed molecule/four-quark states which are in the range $(12326 \sim 12829)$ MeV $(J^P = 1^-)$.
  - We conclude that the masses of the XZ experimental candidates are compatible with (almost) pure $J^{PC} = 1^{+\pm}$, 0^{++} molecule or/and four-quark states. The ones of the $1^{-\pm}$ and $0^{-\pm}$ molecule/four-quark states are about 1.5 GeV above the $Y_{c,b}$ mesons experimental candidates and hadronic thresholds.

- *SU(3) breakings*
  - The 0^{++} $X(4350)$ experimental candidate might be identified with a $\bar{D}_{s0}^*D_{s0}^*$ molecule ground state.
  - The masses of 1^{++} $X(4147)$ and $X(4273)$ are compatible within the error with the one of the $\bar{D}_s^*D_s$ molecule state and with the A_c four-quark state.
  - Our predictions suggest the presence of 0^{++} $\bar{D}_sD_s$ and $\bar{D}_s^*D_s^*$ molecule states in the range $(4121 \sim 4396)$ MeV.
  - In the b-channel, we present new predictions for the $0^{-\pm}$, $1^{-\pm}$ and for different beauty states which can be tested in future experiments.
  - Our results show that the SU3 breakings are relatively small for the masses ≤ 10 (respectively 3)% for the charm (respectively bottom) channels while it can be

large for the couplings ($\leq 20\%$). In the ratios of sum rules, the corrections tend to cancel out.

Acknowledgments

We would like to thank R. M. Albuquerque, S. Narison and G. Randriamanatrika for their valuable contributions to the research summarized in this paper.

ORCID

A. Rabemananjara https://orcid.org/0009-0003-5851-4207
D. Rabetiarivony https://orcid.org/0000-0002-8392-8634

References

1. R. Albuquerque *et al.*, *Int. J. Mod. Phys. A* **31**, 1650196 (2016).
2. R. Albuquerque *et al.*, *Int. J. Mod. Phys. A* **33**, 1850082 (2018).
3. BELLE Collab. (S.-K. Choi *et al.*), *Phys. Rev. Lett.* **91**, 262001 (2003).
4. BESIII Collab. (M. Albikim *et al.*), *Phys. Rev. Lett.* **110**, 252001 (2013).
5. BELLE Collab. (K.-F. Chen *et al.*), *Phys. Rev. Lett.* **100**, 112001 (2008).
6. BELLE Collab. (A. Bondar *et al.*), *Phys. Rev. Lett.* **108**, 122001 (2012).
7. CDF Collab. (T. Aaltonen *et al.*), *Phys. Rev. Lett.* **102**, 242002 (2009).
8. CMS Collab. (S. Chatrchyan *et al.*), *Phys. Lett. B* **734**, 261 (2014).
9. D0 Collab. (V. M. Abazov *et al.*), *Phys. Rev. D* **89**, 012004 (2014).
10. BELLE Collab. (C. P. Shen *et al.*), *Phys. Rev. Lett.* **104**, 112004 (2010).
11. E. S. Swanson, *Phys. Rept.* **429**, 243 (2006); J. M. Richard, *Nucl. Phys. (Proc. Suppl.) B* **164**, 131 (2007); F. S. Navarra, M. Nielsen and S. H. Lee, *Phys. Rep.* **497**, 41 (2010); N. Brambilla *et al.*, *Eur. Phys. J. C* **71**, 1534 (2011); S. L. Zhu, *Int. J. Mod. Phys. E* **17**, 283 (2008).
12. M. A. Shifman, A. I. Vainshtein and V. I. Zakharov, *Nucl. Phys. B* **147**, 385 (1979).
13. S. Narison, *QCD as a Theory of Hadrons*, Vol. 17 (Cambridge Monographs on Particle Physics, Nuclear Physics and Cosmology, 2002), p. 1.
14. S. Narison, *QCD Spectral Sum Rules*, World Scientific Lecture Notes in Physics, Vol. 26 (World Scientific, 1989), p. 1.
15. S. Narison, *Phys. Rept.* **84**, 263 (1982); *Acta Phys. Pol. B* **26**, 687 (1995).
16. S. Narison, *Phys. Lett. B* **210**, 238 (1988).
17. A. Pich and E. de Rafael, *Phys. Lett. B* **158**, 477 (1985).
18. S. Narison and A. Pivovarov, *Phys. Lett. B* **327**, 341 (1994).
19. S. Narison, *Int. J. Mod. Phys. A* **33**, 1850045 (2018).
20. S. Narison, arXiv:1812.09360 [hep-ph].
21. S. Narison, *Int. J. Mod. Phys. A* **30**, 1550116 (2015).
22. S. Narison, *Phys. Lett. B* **738**, 346 (2014).
23. S. Narison, *Phys. Lett. B* **784**, 261 (2018); *Phys. Lett. B* **802**, 135221 (2020).
24. B. L. Ioffe, *Nucl. Phys. B* **188**, 317 (1981).
25. Y. Chung *et al.*, *Z. Phys. C* **25**, 151 (1984).
26. H. G. Dosch, M. Jamin and S. Narison, *Phys. Lett. B* **220**, 251 (1989).
27. B. L. Ioffe, *Nucl. Phys. B* **191**, 591 (1981).
28. A. A. Ovchinnikov and A. A. Pivovarov, *Yad. Fiz.* **48**, 1135 (1988).
29. S. Narison, *Phys. Lett. B* **605**, 319 (2005).

30. S. Narison, *Phys. Lett. B* **693**, 559 (2010) [Erratum-ibid. **705**, 544 (2011)].
31. S. Narison, *Phys. Lett. B* **706**, 412 (2011).
32. S. Narison, *Phys. Lett. B* **707**, 259 (2012).
33. R. A. Bertlmann, G. Launer and E. de Rafael, *Nucl. Phys. B* **250**, 61 (1985).
34. S. Narison, *Phys. Lett. B* **673**, 30 (2009).
35. G. Launer, S. Narison and R. Tarrach, *Z. Phys. C* **26**, 433 (1984).
36. E. G. Floratos, S. Narison and E. de Rafael, *Nucl. Phys. B* **155**, 155 (1979).
37. R. Tarrach, *Nucl. Phys. B* **183**, 384 (1981).
38. R. Coquereaux, *Ann. Phys.* **125**, 401 (1980).
39. P. Binetruy and T. Sücker, *Nucl. Phys. B* **178**, 293 (1981).
40. S. Narison, *Phys. Lett. B* **197**, 405 (1987).
41. S. Narison, *Phys. Lett. B* **216**, 191 (1989).

Structure of the soft theorem toward amplitudes construction*

Roland Raboanary[†,§] and Andriniaina Narindra Rasoanaivo [‡,§,¶]

[†]*Physique et Applications, Faculté des Sciences,*
Université d'Antananarivo-Madagascar

[‡]*Sciences Expérimentales et des Mathématiques, Ecole Normale Supérieure,*
Université d'Antananarivo - Madagascar

[§]*Institute of High-Energy Physics of Madagascar (iHEPMAD),*
University of Antananarivo - Madagascar
[¶]*andriniaina@aims.ac.za*

Received 18 February 2025
Accepted 31 July 2025
Published 10 September 2025

In this paper, we present the importance of soft radiation's contribution for high-precision physics to match the theoretical predictions with the experimental measurements. We briefly review the different properties of soft factorization in the context of scattering amplitude. The universal behavior of the IR contribution in this factorization plays the central focus of studies in order to recover any information about the UV contribution which allows us to develop an algorithm to softly add particles to compute higher point amplitudes.

Keywords: Soft theorem; soft operator; soft decomposition; helicity amplitudes.

1. Introduction

Scattering amplitude is the core of quantum field theory, providing a critical bridge between theoretical predictions and experimental observations. It describes the probabilities of particle interactions, offering valuable insights into the fundamental interactions and the properties of particles that define our universe. By studying amplitudes, we not only test the Standard Model's predictions but also explore new physics that could extend beyond it.

A key aspect of understanding scattering amplitudes is to analyze their behavior in the soft limit, where a particle's momentum approaches zero. The soft theorem, introduced by Weinberg in 1965, describes this behavior[1] and has proven to be indispensable in matching the theoretical predictions to the high-precision

*Talk given at the 14th International Conference in High-Energy Physics - HEPMAD24 (21–26th October 2024, Antananarivo-Madagascar)

¶Corresponding author.

experimental measurements by considering the effects of soft emissions.[2] In heavy ion physics, it is important to consider the contribution of induced soft radiation to fully capture the energy loss of jets through the quark gluon plasma.[3,4] By incorporating the contributions of soft emissions, we can probe the properties of the plasma and unravel the dynamics of QCD interaction in different regimes.

The focus of this work is on studying the properties of soft factorization and the contributions of the different scales in the soft theorem. Specifically, we aim to explore how the soft limit can be inverted to derive higher-point amplitudes from lower points. This approach not only deepens our understanding of the structure of scattering amplitudes but also opens new pathways for constructing amplitudes with complex helicity configurations.

To achieve this, we examine the universal properties of soft factors, their dependence on particle spin and helicity and the decomposition of amplitudes into infrared (IR) and ultraviolet (UV) contributions.[5] By applying these concepts, we investigate specific helicity configurations of gluons, focusing on how higher-point amplitudes can be constructed through the addition of soft particles. This process, referred to as the inverse soft limit,[6,7] highlights the interplay between soft and hard contributions to amplitudes.

This work emphasizes the importance of understanding soft factorization and its role in scattering amplitudes. By exploring the inverse soft limit, we aim to provide new insights into how higher-point amplitudes can be systematically derived, offering a novel perspective on the relationships between different amplitude configurations. In doing so, we hope to further bridge the gap between theoretical developments on inverse soft approaches.

2. Structure of the Soft Theorem

2.1. *Soft operator*

Weinberg's soft theorem is a universal property of scattering amplitudes, in which the soft contribution of the amplitude is factorized out from the high-energy contribution through the relation

$$A_n(p_1,\ldots,\epsilon p_n) \xrightarrow{\epsilon\to 0} S(\epsilon p_n)A_{n-1}(p_1,\ldots,p_{n-1}). \tag{1}$$

The soft contribution, denoted by $S(p_n)$, is the soft contribution and is known as an operator acting on the hard contribution. To achieve our goal of reconstructing amplitudes by inverting the soft calculation, it is important to review the structures and properties of the soft factorization. We first review the properties of soft operators and then describe the structure/mechanism of the soft theorem which is the main focus of this work.

The dimension analysis of the differential cross-section showed that the mass dimension of the kinematics of the scattering amplitude doesn't depend only on the number of particles involved but it depends also on the properties of the particles and the types of interactions involved in the process. On one hand, it is well understood that n-point amplitudes M_n where all n momenta are taken to be outgoing have a physical dimension independent from the nature of the particles. Where in the regular four dimension space–time, the mass dimension of amplitudes is given by

$$[M_n] = 4 - n. \tag{2}$$

On the other hand, we can prove from diagram analysis (Feynman diagrams) that in pure gauge, if g_s is the coupling strength, the n-point tree amplitude can be written as

$$M_n = g_s^{n-2} A_n, \tag{3}$$

up to another dimensionless factor from non-space-time symmetry.[8] With all that in hand, we can find that the mass dimension of a pure kinematic n-point amplitude A_n is given by

$$[A_n] = 4 - n - (n-2)[g_s], \tag{4}$$

where the dimension of the coupling is given by $[g_s] = 1 - s$, which leads to

$$[A_n] = 6 - 2n + (n-2)s. \tag{5}$$

The soft operator is well studied from its expansion in powers of the soft energy. The leading term is a function known as *the Weinberg soft factor* $S^{(0)}$ and it is found to be infrared divergent and appear to be universal[1]; in the case of a photon such is given by

$$S^{(0)}(p) = \left(\frac{k_1 \cdot \varepsilon(p)}{p \cdot k_1} - \frac{k_2 \cdot \varepsilon(p)}{p \cdot k_2}\right). \tag{6}$$

The sub-leading, $S^{(1)}$, term is found to be more complex which involves the contribution of total angular momentum operator $J^{\mu\nu}$ that acts on the hard momentum contribution A_{n-1}. The sub-leading term is mainly derived by considering the momentum deformation of adjacent particles so the energy–momentum of the born amplitude A_{n-1} is conserved[9]; such deformation is mainly produced from the action of $J^{\mu\nu}$ which leads to the following expression:

$$S^{(1)}(p) = \left(\frac{p_\mu \varepsilon_\nu(p) J_1^{\mu\nu}}{p \cdot k_1} - \frac{p_\mu \varepsilon_\nu(p) J_2^{\mu\nu}}{p \cdot k_2}\right). \tag{7}$$

Much progress has been made in studying the different terms in the soft expansion of the soft operator and their universal behavior, and out of these studies the universal re-summation of the soft contribution can be written as a product of the Weinberg soft factor with a unitary operator U that shifts the adjacent momenta to preserve momentum conservation both for n-point amplitude and the born amplitude $(n-1)$,[7]

$$S(p) = S^{(0)}(p) U(k_1, k_2). \tag{8}$$

The action of the unitary operator $U(k_1, k_2)$ on the adjacent momenta k_1 and k_2, as the momentum p get soft, can be illustrated with the following diagram:

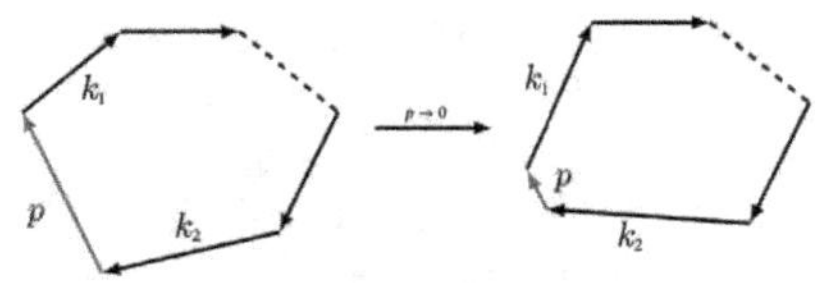

(9)

2.2. *Inverse soft limit*

The inverse soft limit, also known as ISL, is a technique to invert the soft factorization by adding softly particle to a given amplitude to obtain a larger point amplitude .[6,7] Such a mechanism is usually done by applying the soft operator on the lower point amplitude, however, such action doesn't lead to a larger point amplitude. To deepen the understanding of the inverse soft limit, it is important to review the mechanism of the factorization as one of the momentum goes soft. Under the limit, the amplitude behaves similarly to any continuous function $f(x)$ in the asymptotic limit which is dominated by the infrared behavior at low x, while the generic part is the ultraviolet behavior that tends to zero, as an example

$$f(x) = \frac{0.1}{x^n} + 100x^{n+1} \xrightarrow{x \to 0} \frac{0.1}{x^n}. \tag{10}$$

In the soft limit, the amplitude behaves the same way as the above function (10), i.e. the amplitude can be decomposed in two regimes: first, an IR regime which is the main subject of the soft theorem, and second a UV regime which vanish at low momentum.[5]

For the case of MHV amplitudes, which contains only two negative helicities, the action of adding a positive helicity gluon leads to a higher point MHV amplitude. Generalizing a similar idea for any type of amplitude by adding any helicity gluon is then a solution to softly reconstruct higher point amplitude from a lower one. However, such generalization is not trivial because of the structure of the amplitude. The action of the soft operator only involved the IR regime which is the central point of the soft theorem while the UV regime is a reminder function that needs to be found.

3. Road to an Invert Soft Reconstruction

3.1. *Computational search*

To achieve the objective of soft reconstruction amplitude, we mainly focus on the IR/ UV structure of the amplitude to shape the possibility of constructing helicity gluons amplitudes from lower points by inverse soft. We are considering known helicity amplitudes from six to eight points of all possible helicity configurations as computed from Ref. 9. For each amplitude, we set one particle to be softer and then invert the soft limit to identify the IR and UV contribution. Then we do the same process on the same amplitude with a different particle, and so on. This computation is to observe if we can get more information on the soft decomposition.

To illustrate this experimental computation, we will consider the following simplest nontrivial helicity configuration represented using the spinor helicity variables as in Ref. 10

$$\begin{aligned} A_6(1^-,2^-,3^-,4^+,5^+,6^+) = \frac{1}{[2|3+4|5\rangle}\Bigg(&\frac{[4|5+6|1\rangle^3}{[34][23]\langle 56\rangle\langle 61\rangle s_{234}} \\ &+\frac{[6|1+2|3\rangle^3}{[61][12]\langle 34\rangle\langle 45\rangle s_{612}}\Bigg) \end{aligned} \tag{11}$$

and as we take the third particle soft and then identify reminder function

$$A_6 \xrightarrow{p_3\to 0} A_{IR}^{(3)} = \frac{1}{[2|3+4|5\rangle}\left(\frac{[4|5+6|1\rangle^3}{[34][23]\langle 56\rangle\langle 61\rangle s_{234}}\right) \tag{12}$$

This is the asymptotic behavior of the amplitude in the nonfactorized form. From that, it is evident the UV regime

$$A_{UV}^{(3)} = \frac{1}{[2|3+4|5\rangle}\left(\frac{[6|1+2|3\rangle^3}{[61][12]\langle 34\rangle\langle 45\rangle s_{612}}\right). \tag{13}$$

A similar process is done for each one of the six particles to identify them individual IR and UV regimes.

3.2. *Observation*

After applying the same computational process to different helicity configuration amplitudes, six to eight points, we observed a surprising pattern that can be used to compute amplitudes. We found that even if the UV information is not inaccessible via the soft theorem, it is simply hidden in several different soft limits.

For example, by using the amplitude (11), instead of the third particle we can observe its asymptotic behavior according to the sixth particle, which is a positive helicity gluon. We obtain the same decomposition into IR and UV regimes as in (12) and (13). However, by observing the decomposition according to the fourth particle, (12) becomes the UV regime while (13) is the IR.

$$A_{IR}^{(4)} = A_{UV}^{(3)} \quad \text{and} \quad A_{UV}^{(4)} = A_{IR}^{(3)}. \tag{14}$$

Since the action of the soft operator only leads to the IR regime of the amplitude, and with the fact that the UV regime can be obtained from a different soft action, the larger six point amplitude can be obtained from the actions of two soft operators, respectively, on two lower points as follows:

$$\begin{aligned} A_6(1^-,2^-,3^-,4^+,5^+,6^+) &= S(3^-)A_5(1^-,2^-,4^+,5^+,6^+) \\ &\quad + S(4^+)A_5(1^-,2^-,3^-,5^+,6^+). \end{aligned} \tag{15}$$

Similar things appear with different helicity configurations of six point amplitudes. Depending on the configuration, to construct the six amplitude, we sometimes require three soft operators to fully reconstruct the amplitude. This is the case for $A_6(1^-, 2^-, 3^+, 4^-, 5^+, 6^+)$, we need to combine the contribution of three soft operators, respectively, acting on $A_5(1^-, 2^-, 4^-, 5^+, 6^+)$, $A_5(1^-, 2^-, 3^+, 5^+, 6^+)$ and $A_5(1^-, 2^-, 3^+, 4^-, 6^+)$. The same thing can be observed for different configurations for seven and eight point amplitudes where either two or three contributions are needed for the complete reconstruction.

Among the different helicity configurations of the six, seven and eight point amplitudes, a simpler pattern appears for the case of amplitudes where all negative helicity are grouped in the ordering. The ISL algorithm for such simplicity requires only the action of two soft operators as in (17). Those amplitudes are of the following form:

$$A_n(\underbrace{1^-, \ldots, m^-}_{h=-1}, \underbrace{(m+1)^+, \ldots, n^+}_{h=+1}), \tag{16}$$

where the first m gluons have positive helicities. In our case, those amplitudes are as follows: $A_6(1^-, 2^-, 3^-, 4^+, 5^+, 6^+)$, $A_7(1^-, 2^-, 3^-, 4^+, 5^+, 6^+, 7^+)$, $A_7(1^-, 2^-, 3^-, 4^-, 5^+, 6^+, 7^+)$, $A_8(1^-, 2^-, 3^-, 4^-, 5^+, 6^+, 7^+, 8^+)$, $A_8(1^-, 2^-, 3^-, 4^+, 5^+, 6^+, 7^+, 8^+)$, $A_8(1^-, 2^-, 3^-, 4^-, 5^-, 6^+, 7^+, 8^+)$ and them cyclic helicity permutation configurations.

We attempt to generalize the pattern (17) for any n-point amplitude of the form (16). We use the Britto Cachazo Fen Witten recursion ,[9] by identifying the BCFW shifts with the soft momentum shifts as in (8). Details related to the proof are the subject of another paper which will show the following result:

$$\begin{aligned} &A_n(1^-, \ldots, m^-, (m+1)^+, \ldots, n^+) \\ &\quad = S(m^-)A_{n-1}(1^-, \ldots, (m-1)m, (m+1)^+, \ldots, n^+) \\ &\quad + S((m+1)^+)A_{n-1}(1^-, \ldots, m^-, (m+2)^+, \ldots, n^+). \end{aligned} \tag{17}$$

4. Conclusion

A deep understanding of scattering amplitudes is fundamental to unraveling the hidden properties in the experimental measurements. This work demonstrates the construction of higher-point amplitudes by systematically incorporating soft gluons through the soft decomposition approach into IR and UV regimes. We successfully derive these amplitudes by focusing on specific helicity configurations, highlighting the intricate interplay between soft and hard contributions. This achievement underscores the utility of soft factorization as a tool for systematically building higher point scattering amplitudes and deepens our understanding of their underlying structure. The results pave the way for further exploration of the soft limit and its applications.

ORCID

Andriniaina Narindra Rasoanaivo https://orcid.org/0000-0002-4842-9741

References

1. S. Weinberg, *Phys. Rev.* **140**, B516 (1965).
2. L. Flower and M. Schoenherr, *J. High Energy Phys.* **03**, 238 (2023).
3. A. N. Rasoanaivo and W. A. Horowitz, arXiv:1712.06292 [hep-ph].
4. W. A. Horowitz, *Nucl. Phys. A* **904–905**, 186c (2013).
5. L. Rodina, *Phys. Rev. Lett.* **122**, 071601 (2019).
6. D. Nandan and C. Wen, *J. High Energy Phys.* **08**, 040 (2012).
7. C. Boucher-Veronneau and A. J. Larkoski, *J. High Energy Phys.* **09**, 130 (2011).
8. Z. Bern and D. A. Kosower, *Nucl. Phys. B* **362**, 389 (1991).
9. T. McLoughlin, A. Puhm and A. M. Raclariu, *J. Phys. A* **55** (2022) 443012.
10. R. Britto, F. Cachazo and B. Feng, *Nucl. Phys. B* **715**, 499 (2005).
11. S. Badger, J. Henn, J. C. Plefka and S. Zoia, *Scattering Amplitudes in Quantum Field Theory, Lecture Notes in Physics*, Vol. 1021 (Springer, 2024).

Part 4

Beyond the Standard Model and Neutrinos Physics[4]

Chairs: R. Raboanary, C. Ratsifaritana

[4]These articles will not be included in this publication due to copyright restrictions.

Search for Long-Lived Charged Particles Using the CMS Detector in Run-2

T.A. Vami (UC-Santa-Barbara, US)

Weyl Anomaly in the Standard Model

P. Dominis Prester (Univ. Rijeka, Croatia)

Using Graph Neural Networks for Finding and Reconstructing Neutrinos

T.C Petersen (Univ. Copenhagen, DK)

Quantum Phase Space Symmetry and Sterile Neutrinos

R.T. Ranaivoson (INSTN-Antananarivo, MG)

Part 5

Astro and Solar Physics

Chairs: T. Fanjirinjatovo, R. Raboanary,
C. Ratsifaritana

Atmosphere in VHE Gamma-ray Astronomy[5]

J. Dominis Prester (Univ. Rijeka, Croatia)

[5] Manuscript not received.

Gravitational collapse with torsion and universe in a black hole*

Nikodem Popławski

Department of Mathematics and Physics,
University of New Haven
300 Boston Post Road, West Haven, CT 06516, USA
NPoplawski@newhaven.edu

Received 21 February 2025
Accepted 31 July 2025
Published 17 September 2025

We consider gravitational collapse of a fluid sphere with torsion generated by spin, which forms a black hole. We use the Tolman metric and the Einstein–Cartan field equations with a relativistic spin fluid as a source. We show that gravitational repulsion of torsion prevents a singularity, replacing it with a nonsingular bounce. Quantum particle creation during contraction prevents shear from overcoming torsion. Particle creation during expansion can generate a finite period of inflation and produce large amounts of matter. The resulting closed universe on the other side of the event horizon may have several bounces. Such a universe is oscillatory, with each cycle larger than the preceding cycle, until it reaches a size at which dark energy dominates and expands indefinitely. Our Universe might have therefore originated from a black hole existing in another universe.

Keywords: Einstein–Cartan theory; torsion; spin; gravitational collapse; regular black hole; big bounce.

1. Einstein–Cartan Gravity and Spin Fluid

In General Relativity (GR), the affine connection $\Gamma^{\,k}_{ij}$ is symmetric.[1] The Einstein–Cartan–Sciama–Kibble (EC) theory gravity removes this constraint by regarding the antisymmetric part of the connection, the torsion tensor $S^k_{\ ij} = (1/2)(\Gamma^{\,k}_{ij} - \Gamma^{\,k}_{ji})$, as a field. The total Lagrangian density is $(-1/2\kappa)R\sqrt{-g} + \mathcal{L}_m$, as in GR, where R is the Ricci scalar constructed from the connection $\Gamma^{\,k}_{ij}$, $\mathcal{L}_m$ is the Lagrangian density of matter, and g is the determinant of the metric tensor g_{ik}.

Varying the Lagrangian with respect to the contortion tensor $C^k_{\ ij} = S_{ij}^{\ \ k} + S_{ji}^{\ \ k} + S^k_{\ ij}$ gives the Cartan equations[2]:

$$S_{jik} - S_i g_{jk} + S_k g_{ji} = -\frac{1}{2}\kappa s_{ikj},$$

*Talk given at the 14th International Conference in High-Energy Physics — HEPMAD24 (21–26th October 2024, Antananarivo-Madagascar)

where $S_i = S^k{}_{ik}$ and $s^{ijk} = 2(\delta\mathcal{L}_m/\delta C_{ijk})/\sqrt{-g}$ is the spin tensor of matter. Varying the Lagrangian with respect to the metric tensor g_{ik} gives the Einstein equations with the Ricci tensor constructed from Γ^k_{ij}. They can be put into a GR form with the Einstein tensor G^{ik} and the energy–momentum tensor of matter T^{ik}, combined with contributions from the spin tensor:

$$G^{ik} = \kappa T^{ik} + \frac{1}{2}\kappa^2\Big(s^{ij}{}_j s^{kl}{}_l - s^{ij}{}_l s^{kl}{}_j - s^{ijl}s^k{}_{jl} + \frac{1}{2}s^{jli}s_{jl}{}^k$$
$$+ \frac{1}{4}g^{ik}(2s^{jlm}s_{jml} - 2s_{jl}{}^l s^{jm}{}_m + s^{jlm}s_{jlm})\Big).$$

Dirac spinors, representing fermions, couple to torsion through the covariant derivative in the Lagrangian, so their intrinsic angular momentum (spin) is the source of torsion. At macroscopic scales, they can be averaged and described as a spin fluid:

$$s_{ij}{}^k = s_{ij}u^k, \quad s_{ij}u^j = 0, \quad s^2 = \frac{1}{2}s_{ij}s^{ij} \propto n_f^2 > 0,$$

where s^2 is the averaged square of the spin density and n_f is the number density of fermions. The terms in the combined energy–momentum tensor, which are quadratic in the spin tensor, do not vanish after averaging:

$$G^{ij} = \kappa\Big(\epsilon - \frac{1}{4}\kappa s^2\Big)u^i u^j - \kappa\Big(p - \frac{1}{4}\kappa s^2\Big)(g^{ij} - u^i u^j).$$

The Einstein–Cartan equations for a spin fluid are therefore equivalent to the Einstein equations in GR for an ideal fluid with

$$\tilde{\epsilon} = \epsilon - \alpha n_f^2, \quad \tilde{p} = p - \alpha n_f^2, \tag{1}$$

where ϵ and p are the thermodynamic energy density and pressure, and $\alpha = \kappa(\hbar c)^2/32$ with $\kappa = 8\pi G/c^4$.[3–5]

2. Gravitational Collapse of a Fluid Sphere

A centrally symmetric gravitational field is given by the Tolman metric[6]:

$$ds^2 = e^{\nu(\tau,R)}c^2d\tau^2 - e^{\lambda(\tau,R)}dR^2 - e^{\mu(\tau,R)}(d\theta^2 + \sin^2\theta d\phi^2), \tag{2}$$

where ν, λ and μ are functions of a time coordinate τ and a radial coordinate R. Coordinate transformations $\tau \to \tilde{\tau}(\tau)$ and $R \to \tilde{R}(R)$ do not change the form of this metric. The components of the Einstein tensor corresponding to the metric (2) that do not vanish identically are[6]

$$G^0_0 = -e^{-\lambda}\Big(\mu'' + \frac{3}{4}\mu'^2 - \frac{1}{2}\mu'\lambda'\Big) + \frac{1}{2}e^{-\nu}\Big(\dot{\lambda}\dot{\mu} + \frac{1}{2}\dot{\mu}^2\Big) + e^{-\mu},$$
$$G^1_1 = -\frac{1}{2}e^{-\lambda}\Big(\frac{1}{2}\mu'^2 + \mu'\nu'\Big) + e^{-\nu}\Big(\ddot{\mu} - \frac{1}{2}\dot{\mu}\dot{\nu} + \frac{3}{4}\dot{\mu}^2\Big) + e^{-\mu},$$

$$
\begin{aligned}
G_2^2 = G_3^3 = &-\frac{1}{4}e^{-\lambda}(2\nu'' + \nu'^2 + 2\mu'' + \mu'^2 - \mu'\lambda' - \nu'\lambda' + \mu'\nu') \\
&-\frac{1}{4}e^{-\nu}(\dot{\lambda}\dot{\nu} + \dot{\mu}\dot{\nu} - \dot{\lambda}\dot{\mu} - 2\ddot{\lambda} - \dot{\lambda}^2 - 2\ddot{\mu} - \dot{\mu}^2), \\
G_0^1 = &\frac{1}{2}e^{-\lambda}(2\dot{\mu}' + \dot{\mu}\mu' - \dot{\lambda}\mu' - \dot{\mu}\nu'),
\end{aligned}
\tag{3}
$$

where a dot denotes differentiation with respect to $c\tau$ and a prime denotes differentiation with respect to R.

In the comoving frame of reference, the spatial components of the four-velocity u^i vanish. The nonzero components of the energy–momentum tensor for a spin fluid, $T_{ik} = (\tilde{\epsilon} + \tilde{p})u_i u_k - \tilde{p}g_{ik}$, are: $T_0^0 = \tilde{\epsilon}$, $T_1^1 = T_2^2 = T_3^3 = -\tilde{p}$. The Einstein field equations $G_k^i = \kappa T_k^i$ in this frame of reference are: $G_0^0 = \kappa\tilde{\epsilon}$, $G_1^1 = G_2^2 = G_3^3 = -\kappa\tilde{p}$, $G_0^1 = 0$. The covariant conservation of the energy–momentum tensor gives[6]

$$
\dot{\lambda} + 2\dot{\mu} = -\frac{2\dot{\tilde{\epsilon}}}{\tilde{\epsilon} + \tilde{p}}, \quad \nu' = -\frac{2\tilde{p}'}{\tilde{\epsilon} + \tilde{p}},
\tag{4}
$$

where the constants of integration depend on the allowed transformations $\tau \to \tilde{\tau}(\tau)$ and $R \to \tilde{R}(R)$.

If the pressure is homogeneous (no pressure gradients), then $\tilde{p}' = 0$ and $p = p(\tau)$. In this case, the second equation in (4) gives $\nu' = 0$. Therefore, $\nu = \nu(\tau)$ and a transformation $\tau \to \tilde{\tau}(\tau)$ can bring ν to zero and $g_{00} = e^{\nu}$ to 1. The system of coordinates becomes synchronous. Defining $r(\tau, R) = e^{\mu/2}$ turns the metric (2) into

$$
ds^2 = c^2 d\tau^2 - e^{\lambda(\tau,R)}dR^2 - r^2(\tau,R)(d\theta^2 + \sin^2\theta\, d\phi^2).
\tag{5}
$$

The Einstein equations (3) reduce to[6]

$$
\begin{aligned}
\kappa\tilde{\epsilon} = &-\frac{e^{-\lambda}}{r^2}(2rr'' + r'^2 - rr'\lambda') + \frac{1}{r^2}(r\dot{r}\dot{\lambda} + \dot{r}^2 + 1), \\
-\kappa\tilde{p} = &\frac{1}{r^2}(-e^{-\lambda}r'^2 + 2r\ddot{r} + \dot{r}^2 + 1), \\
-2\kappa\tilde{p} = &-\frac{e^{-\lambda}}{r}(2r'' - r'\lambda') + \frac{\dot{r}\dot{\lambda}}{r} + \ddot{\lambda} + \frac{1}{2}\dot{\lambda}^2 + \frac{2\ddot{r}}{r}, \\
2\dot{r}' - \dot{\lambda}r' = &\, 0.
\end{aligned}
\tag{6}
$$

Integrating the last equation in (6) gives

$$
e^{\lambda} = \frac{r'^2}{1 + f(R)},
\tag{7}
$$

where f is a function of R satisfying a condition $1 + f > 0$.[6] Putting the relation (7) into the second equation in (6) gives $2r\ddot{r} + \dot{r}^2 - f = -\kappa\tilde{p}r^2$, which is integrated to[7]

$$
\dot{r}^2 = f(R) + \frac{F(R)}{r} - \frac{\kappa}{r}\int \tilde{p}r^2 dr,
\tag{8}
$$

where F is a positive function of R. The third equation in (6) does not give a new relation. Putting the relation (7) into the first equation in (6) and using equation (8) gives $\kappa(\tilde{\epsilon}+\tilde{p}) = F'(R)/(r^2 r')$.[7] Combining it with Eq. (8) gives

$$\dot{r}^2 = f(R) + \frac{\kappa}{r}\int_0^R \tilde{\epsilon} r^2 r' dR. \tag{9}$$

Every particle in a collapsing fluid sphere is represented by a radial coordinate R that ranges from 0 (at the center of the sphere) to R_0 (at the surface of the sphere). The relation (9) is the equation of radial motion for a particle with a given value of R. If the mass of the sphere is M, then the Schwarzschild radius $r_g = 2GM/c^2$ of the black hole that forms from the sphere is equal to $r_g = \kappa\int_0^{R_0} \tilde{\epsilon} r^2 r' dR$.[6] Equation (9) for $R = R_0$ gives the equation of motion for a particle at the surface of the sphere:

$$\dot{r}^2(\tau, R_0) = f(R_0) + \frac{r_g}{r(\tau, R_0)}. \tag{10}$$

If $r_0 = r(0, R_0)$ is the initial radius of the sphere and the sphere is initially at rest, then $\dot{r}(0, R_0) = 0$. Consequently, the relation (10) determines the value of R_0:

$$f(R_0) = -\frac{r_g}{r_0}. \tag{11}$$

Putting the relations $r = e^{\mu/2}$ and (7) into the first equation in (4) gives $(\tilde{\epsilon} r^2 r')^{\cdot} + \tilde{p}(r^2 r')^{\cdot} = 0$, which is the first law of thermodynamics for the effective $\tilde{\epsilon}$ and $\tilde{p}$ (1).[4]

3. Collapse of a Spin Fluid Sphere

If we consider a spin fluid, which is composed by an ultrarelativistic matter in kinetic equilibrium, then $\epsilon = h_\star T^4$, $p = \epsilon/3$, and $n_f = h_{nf} T^3$, where T is the temperature of the fluid, $h_\star = (\pi^2/30)(g_b + (7/8)g_f)k_B^4/(\hbar c)^3$, and $h_{nf} = (\zeta(3)/\pi^2)(3/4)g_f k_B^3/(\hbar c)^3$.[4,5] For standard-model particles, $g_b = 29$ and $g_f = 90$. Therefore, the effective energy density and pressure (1) depend on the temperature according to

$$\tilde{\epsilon} = h_* T^4 - \alpha h_{nf}^2 T^6, \quad \tilde{p} = \frac{1}{3} h_* T^4 - \alpha h_{nf}^2 T^6.$$

If the pressure has no gradient, then the temperature depends only on τ, and so does the energy density. This case describes a homogeneous sphere. Putting them into the first law of thermodynamics gives

$$r^2 r' T^3 = g(R), \tag{12}$$

where g is a function of R. Putting this relation into Eq. (9) gives[7]

$$\dot{r}^2 = f(R) + \frac{\kappa}{r}(h_\star T^4 - \alpha h_{nf}^2 T^6)\int_0^R r^2 r' dR. \tag{13}$$

Equations (12) and (13) give the function $r(\tau, R)$, which with the relation (7) gives the function $\lambda(\tau, R)$. The integral of the equation of motion (13) also contains the initial value $\tau_0(R)$. A general solution of the Einstein equations (6) for the metric (5) therefore depends on three arbitrary functions: $f(R)$, $g(R)$, and $\tau_0(R)$.

We seek a solution of Eqs. (12) and (13) as

$$f(R) = -\sin^2 R, \quad r(\tau, R) = a(\tau)\sin R, \tag{14}$$

where $a(\tau)$ is a nonnegative function of τ. For these functions, the relation (12) gives $a^3T^3\sin^2 R\cos R = g(R)$, in which separation of the variables τ and R leads to

$$g(R) = \text{const}\cdot\sin^2 R\cos R, \quad a^3T^3 = \text{const}.$$

Consequently, we obtain

$$aT = a_0T_0, \quad \frac{\dot{T}}{T} + \frac{H}{c} = 0, \tag{15}$$

where $a_0 = a(0)$, $T_0 = T(0)$, and $H = c\dot{a}/a$. Putting the relations (14) into the equation of motion (13) gives[7]

$$\dot{a}^2 + 1 = \frac{1}{3}\kappa(h_\star T^4 - \alpha h_{nf}^2 T^6)a^2, \tag{16}$$

which has the form of the Friedmann equation for the scale factor a as a function of the cosmic time τ in a closed, homogeneous, and isotropic universe. The quantity H is the Hubble parameter of this universe. Using the first relation in (15) to replace T with a in the Friedmann equation (16) yields

$$\dot{a}^2 = -1 + \frac{1}{3}\kappa\left(\frac{h_\star T_0^4 a_0^4}{a^2} - \frac{\alpha h_{nf}^2 T_0^6 a_0^6}{a^4}\right). \tag{17}$$

The values of a_0 and R_0 are determined by the relations (11) and (14)[6,7]:

$$\sin R_0 = \left(\frac{r_g}{r_0}\right)^{1/2}, \quad a_0 = \left(\frac{r_0^3}{r_g}\right)^{1/2}. \tag{18}$$

Putting the initial values a_0 and $\dot{a}(0) = 0$ into Eq. (16), in which the second term on the right-hand side is negligible, gives $Mc^2 = (4\pi/3)r_0^3 h_\star T_0^4$. This relation indicates the equivalence of mass and energy of a fluid sphere with radius r_0 and determines T_0. An event horizon for the sphere forms when $r(\tau, R_0) = r_g$, which is equivalent to $a = (r_g r_0)^{1/2}$. Equation (17) has two turning points, $\dot{a} = 0$, if $r_0^3/r_g > (3\pi G\hbar^4 h_{nf}^4)/(8h_\star^3) \sim l_P^2$ (l_P = Planck length), valid for systems forming black holes.[5]

Putting the relations (14) into the function (7) gives $e^{\lambda(\tau,R)} = a^2$. Consequently, the square of an infinitesimal interval in the interior of a collapsing spin fluid (5) is given by $ds^2 = c^2d\tau^2 - a^2(\tau)dR^2 - a^2(\tau)\sin^2 R(d\theta^2 + \sin^2\theta\, d\phi^2)$[6]. The initial value of the scale factor a is equal to a_0. This metric has a form of the closed Friedmann–Lemaître–Robertson–Walker metric and describes a part of a closed universe with $0 \le R \le R_0$.

4. Nonsingular Bounce and Particle Production

Equation (17) can be solved analytically in terms of an elliptic integral of the second kind,[5] giving the function $a(\tau)$ and then $r(\tau, R) = a(\tau)\sin R$. The value of a never reaches zero because as a decreases, the right-hand side of Eq. (17) becomes negative, contradicting the left-hand side. All particles with $R > 0$ fall within the event horizon but never reach $r = 0$. A singularity is therefore avoided, and replaced with a nonsingular bounce.[7] Positive values of a give finite values of T, ϵ, p and n_f.

After a bounce, the matter expands on the other side of the event horizon as a new, closed universe (with constant positive curvature). The quantity $a(\tau)$ is its scale factor. This universe is oscillatory: a oscillates between the two turning points. The value of R_0 does not change. A turning point at which $\ddot{a} > 0$ is a bounce, and a turning point at which $\ddot{a} < 0$ is a crunch. The universe has therefore an infinite number of cycles (a cycle lasts from a bounce to a crunch, and back).

The Raychaudhuri equation for a congruence of geodesics without four-acceleration and rotation is $d\theta/ds = -\theta^2/3 - 2\sigma^2 - R_{ik}u^i u^k$, where θ is the expansion scalar and σ^2 is the shear scalar. For a spin fluid, the last term in this equation is $-\kappa(\tilde{\epsilon} + 3\tilde{p})/2$. The necessary condition for avoiding a singularity in a black hole is thus $-\kappa(\tilde{\epsilon} + 3\tilde{p})/2 > 2\sigma^2$. For a relativistic spin fluid, $p = \epsilon/3$, this condition gives to $2\kappa\alpha n_f^2 > 2\sigma^2 + \kappa\epsilon$. Without torsion, the left-hand side of this relation would be absent and this inequality could not be satisfied, resulting in a singularity. Torsion therefore provides a necessary condition for preventing a singularity.[8] Without shear, this condition is also sufficient.[3]

The presence of shear opposes the effects of torsion. The shear scalar σ^2 grows with decreasing a like $\sim a^{-6}$, like n_f^2. Therefore, if the initial shear term dominates over the initial torsion term, then it will dominate at later times during contraction, and a singularity would form. To avoid it, n_f^2 must grow faster than $\sim a^{-6}$. Consequently, fermions must be produced in a black hole during contraction.

A quantum production rate of particles in a varying gravitational field is given by

$$\frac{1}{c\sqrt{-g}}\frac{d(\sqrt{-g}n_f)}{d\tau} = \frac{\beta H^4}{c^4}, \quad g = -a^6 \sin^4 R \sin^2\theta, \tag{19}$$

where β is a nondimensional production rate. The second equation in (15) is generalized into[4]

$$\frac{\dot{T}}{T} = \frac{H}{c}\left(\frac{\beta H^3}{3c^3 h_{nf} T^3} - 1\right). \tag{20}$$

Particle production changes the power law $n_f(a)$:

$$n_f \sim a^{-(3+\delta)},$$

where δ varies with τ. Putting this relation into the production rate (19) gives $\delta \sim -a^\delta \dot{a}^3$. During contraction, $\dot{a} < 0$ and thus $\delta > 0$. The term $n_f^2 \sim a^{-6-2\delta}$ grows faster than $\sigma^2 \sim a^{-6}$ and a singularity is avoided. Torsion and particle production together reverse the effects of shear, generating a nonsingular bounce.[7]

5. Closed Universe in a Black Hole and Inflation

The dynamics of the nonsingular, relativistic universe in a black hole (on the other side of its event horizon) is described by Eqs. (16) and (20), with the initial conditions $a(0) = (r_0^3/r_g)^{1/2}$ and $\dot{a}(0) = 0$, giving the functions $a(\tau)$ and $T(\tau)$. The shear appears in Eq. (16) as an additional, positive term $\sim a^{-4}$. When the universe becomes nonrelativistic, the term $h_\star T^4$ changes into a positive term $\sim a^{-1}$. The cosmological constant appears as a positive term $\sim a^2$.

Particle production increases the maximum size of the scale factor a that is reached at a crunch. A new cycle therefore lasts longer than the previous cycle. According to the relations (18), R_0 is given by $\sin^3 R_0 = r_g/a(0)$, where $a(0)$ is the initial scale factor, equal to the maximum a in the first cycle. Because the maximum a in the next cycle is larger, $\sin R_0$ decreases and $R_0 \to \pi$ (completely closed universe)[7].

During contraction, H is negative and the temperature T increases. During expansion, H is positive, so if β is too big, then the right-hand side of equation (20) could become positive. The temperature would then grow with increasing a (eternal inflation). To avoid it, the production rate has an upper limit: The maximum of the positive function $(\beta H^3)/(3c^3 h_{nf} T^3)$ must be smaller than 1.[4]

If the function $(\beta H^3)/(3c^3 h_{nf} T^3)$ increases after a bounce to a value slightly smaller than 1, then T becomes approximately constant. Accordingly, H is also nearly constant and a grows exponentially, generating inflation.[4] Because ϵ during inflation is also nearly constant, the universe produces large amounts of matter and entropy. Such an expansion lasts until this function drops below 1. Consequently, inflation generated by torsion and particle production lasts a finite period of time, after which torsion weakens and the universe begins the radiation- and then the matter-dominated expansion.

If quantum effects near a bounce do not produce enough matter, then the closed universe reaches a maximum size and then contracts to another bounce, beginning a new cycle. Because of matter production, a new cycle reaches a larger maximum size than the previous cycle.[4] When the universe reaches a size at which the cosmological constant dominantes, it avoids another contraction and expands to infinity. The last bounce: the big bounce, is the big bang of the universe created by a black hole.

A parent black hole creating a new, baby universe becomes an Einstein—Rosen bridge (unidirectional wormhole) to that universe.[9] A closed universe is the three-dimensional hypersurface (with radius a) of a four-dimensional hypersphere. Accordingly, our Universe may be closed and born in the interior of a black hole existing in a parent universe[9]. Torsion and particle production generate finite inflation, which is consistent with the astronomical data.[10] This scenario also occurs if the fermionic matter is described by the Dirac fields instead of the spin fluid.[11] It would still be valid for a more realistic gravitational collapse of an inhomogeneous and rotating fluid.

In addition to eliminating gravitational singularities, torsion may also remove divergences in Feynman diagrams in quantum electrodynamics, resulting in finite values of bare (before renormalization) quantities such as the mass and electric charge of the electron.[12] Torsion may be a physical mechanism, ensuring that all physics is finite.

Acknowledgments

I am grateful to Francisco Guedes and my Parents, Bożenna Popławska and Janusz Popławski, for their support.

ORCID

Nikodem Popławski https://orcid.org/0000-0001-5593-313X

References

1. E. Schrödinger, *Space-time Structure* (Cambridge University Press, 1954).
2. T. W. B. Kibble, J. Math. Phys. **2**, 212 (1961); D. W. Sciama, *Rev. Mod. Phys.* **36**, 463 (1964); *Rev. Mod. Phys.* **36**, 1103 (1964); F. W. Hehl, P. von der Heyde, G. D. Kerlick, and J. M. Nester, *Rev. Mod. Phys.* **48**, 393 (1976); E. A. Lord, *Tensors, Relativity and Cosmology* (McGraw-Hill, 1976); N. Popławski, *Classical Physics: Spacetime and Fields*, arXiv:0911.0334; F. R. Benard Guedes and N. J. Popławski, *Class. Quantum Grav.* **41**, 065011 (2024).
3. F. W. Hehl, P. von der Heyde and G. D. Kerlick, Phys. Rev. D **10**, 1066 (1974); B. Kuchowicz, *Gen. Relativ. Gravit.* **9**, 511 (1978); M. Gasperini, *Phys. Rev. Lett.* **56**, 2873 (1986); N. J. Popławski, *Phys. Lett. B* **694**, 181 (2010); *Phys. Lett. B* **701**, 672 (2011); *Gen. Relativ. Gravit.* **44**, 1007 (2012).
4. N. Popławski, *Astrophys. J.* **832**, 96 (2016).
5. G. Unger and N. Popławski, *Astrophys. J.* **870**, 78 (2019).
6. G. Lemaître, *Ann. Soc. Sci. Brux. A* **53**, 51 (1933); R. C. Tolman, *Proc. Natl. Acad. Sci. USA* **20**, 169 (1934); J. R. Oppenheimer and H. Snyder, *Phys. Rev.* **56**, 455 (1939); H. Bondi, *Mon. Not. R. Astron. Soc.* **107**, 410 (1947); L. D. Landau and E. M. Lifshitz, *The Classical Theory of Fields* (Pergamon, 1975).
7. N. Popławski, *Zh. Eksp. Teor. Fiz.* **159**, 448 (2021); *J. Exp. Theor. Phys.* **132**, 374 (2021).
8. W. Kopczyński, *Phys. Lett. A* **39**, 219 (1972); *Phys. Lett. A* **43**, 63 (1973); A. Trautman, *Nature Phys. Sci.* **242**, 7 (1973); F. W. Hehl, P. von der Heyde and G. D. Kerlick, *Phys. Rev. D* **10**, 1066 (1974).
9. R. K. Pathria, *Nature* **240**, 298 (1972); N. J. Popławski, *Phys. Lett. B* **687**, 110 (2010).
10. S. Desai and N. J. Popławski, *Phys. Lett. B* **755**, 183 (2016).
11. N. Popławski, *Phys. Rev. D* **85**, 107502 (2012); J. Magueijo, T. G. Zlosnik and T. W. B. Kibble, *Phys. Rev. D* **87**, 063504 (2013); J. L. Cubero and N. J. Popławski, *Class. Quantum Grav.* **37**, 025011 (2020).
12. N. Popławski, *Found. Phys.* **50**, 900 (2020); M. Del Grosso and N. Popławski, *Class. Quantum Grav.* **41**, 225001 (2024).

Solar radiation mapping for the south west of Madagascar

T. Fanjirindratovo, J. F. Ramaroson* and O. Ramiarinjanahary
Physics and Environment Laboratory, University of Toliara,
P.O. Box 185, Toliara 601, Madagascar
**ramarosonflorence@gmail.com*

Received 24 February 2025
Accepted 31 July 2025
Published 27 September 2025

Mapping solar radiation is essential for optimum management of solar resources. Drawing up solar maps requires reliable knowledge of the spatial and temporal distribution of solar radiation. For this reason, we have used satellite data (Direct Normal Irrradiance) from 1999 to 2003 from METEOSAT-EAST (SARAH-E). Matlab tools were used to extract and process the raw data from for region of interest wich the Atsimo Andrefana region, Madagascar. The main mathematical model is the descriptive statistics to obtain average hourly, daily, monthly and annual values for the solar irradiance. Interpolation techniques were used to draw up maps, showing the spatial distribution of solar radiation by ArcGIS software. After studying the dispersion trend, we noticed that this region is homogeneous in terms of solar solar potential. Thanks to the correlation study between solar irradiation and time, it was found that temporal variation plays a major role in solar irradiance variation.

Keywords: mapping; Direct Normal Irradiance; descriptive statistics; polynomial regression model; correlation; interpolation; Atsimo Andrefana region.

1. Introduction

Solar radiation mapping is essential for optimum management of solar resources. Solar maps are produced either from ground radiometric measurements or from satellite data.[7] Solar radiation data directly affect the solar potential of a given site. Accurate knowledge of the distribution of solar radiation in a given geographic area proves to be of major importance for the development of solar devices,[4] so thermal solar energy is one of the most accessible an sustainable solutions for heat production.[2] The Atsimo Andrefana region is rich in solar deposits but the field is still under-exploited even though research has already been carried out in this area and there are already a number of companies operating in the sector. In this work, we propose the descriptive statistical method as a mathematical model to elaborate temporal and spatial maps of direct solar radiation as a mathematical model for constructing temporal and spatial maps of direct solar radiation.

*Corresponding author.

2. Data Description

The data we use are satellite data from the 'METEOSAT-EAST' (SARAH-E) (geostationary satellite: 'http://www.cmsaf.eu/') with the Network Common Data Form (NETCDF) extension, which is frequently used by many researchers in the fields of climatology and meteorology thanks to its somewhat special characteristics: it can contain voluminous data but is small in size, so easy to transport. Global radiation on a surface is the sum of the three components : direct, diffuse, and the part reflected by the ground.[1] Direct radiation at normal incidence (DNI) is the measurement of radiation arriving by a direct path from the sun on a plane perpendicular to the direction of the ray.[3] We have data stored in daily files measured per unit hour in W/m2 for Madagascar from 1999 to 2013. We can say that the measurement is of high precision because we have a resolution of 5 km × 5 km. The data we have are primary data, i.e. directly collected but not yet analyzed, requiring further processing before use. The raw data we have are spatio-temporal, i.e. they are marked according to latitude, longitude and time.

3. Results and Discussion

In this work, we have chosen descriptive statistics to obtain the maximum information on our data so that we can draw up solar maps according to the situation and we have also used the method of interpolation of values using ArcGIS software to produce spatial solar maps.

3.1. *Descriptive statistics*

Descriptive statistics is the description of each statistical character one by one and not the possible links between them. It is a set of tools and methods for synthesizing and summarizing large volumes of data and large matrices of information.

(a) Average:

$$\bar{X} = \frac{1}{n}\sum_{i=1}^{n} X_i. \tag{1}$$

(b) Variance:

$$\text{var}(X) = \left[\frac{1}{n}\sum_{i=1}^{n} n_i (X_i)^2\right] - (\bar{X})^2. \tag{2}$$

(c) Standard deviation:

$$S_x = \sqrt{\text{var}(X)}. \tag{3}$$

Using the descriptive statistical method and the Matlab tool, we were able to draw up tables showing average values and standard deviations, minimums and maximums of solar radiation for the Atsimo Andrefana region from 1999 to 2003. These mean values show that the region really has a huge potential for solar

Table 1. Average monthly irradiance values for the period 1999–2003.

Month	Average (W/m^2)	Standart deviation (W/m^2)	Minimum (W/m^2)	Maximum (W/m^2)
January	484	83.4	405	623
February	503	37.51	470	546
March	537	37.82	489	577
April	554	16.86	535	576
May	517	32.9	460	538
June	521	26.35	480	545
July	517	25.5	488	555
August	533	23.48	506	565
September	561	31.18	537	602
October	574	20.3	546	602
November	550	52.83	460	589
December	523	33.56	483	559

Table 2. Spatial variation of solar radiation in the Atsimo Andrefana region.

District	Average DNI January W/m^2	Average DNI October W/m^2
Toliara I	546	593
Toliara II	485	586
Sakaraha	430	558
Ankazoabo	438	573
Beroroha	498	566
Betioky-Sud	503	579
Benenitra	488	575
Ampanihy	525	552
Morombe	436	587

radiation, confirming our hypothesis in the introduction. The standard deviations indicate the variation in values. We can see that, even if the area has a lot of sunshine, there are always variations over time and between districts. Here we have two examples of tables illustrating what we have mentioned above. Table 1 presents the monthly average of solar radiation, along with the standard deviations as well as the minimum and maximum values, for the period from 1999 to 2003. Table 2 was prepared in order to compare the solar radiation values between the months of October and January. The Atsimo Andrefana region is made up of nine districts, including Toliara I, Toliara II, Sakaraha, Ampanihy, South-Betioky, Benenitr, Ankazoabo, Beroroha and Morombe. Figure 1 illustrates this.

3.2. *The polynomial regression and the coefficient of interrelationship*

3.2.1. *The polynomial regression*

A polynomial regression model is a non-linear regression model with $f(t_1, \dots, t_p, \beta_0, \dots, \beta_p)$ being a polynomial in $t_1, \dots, t_p$ and coefficients $\beta_0, \dots, \beta_p$. Of generic

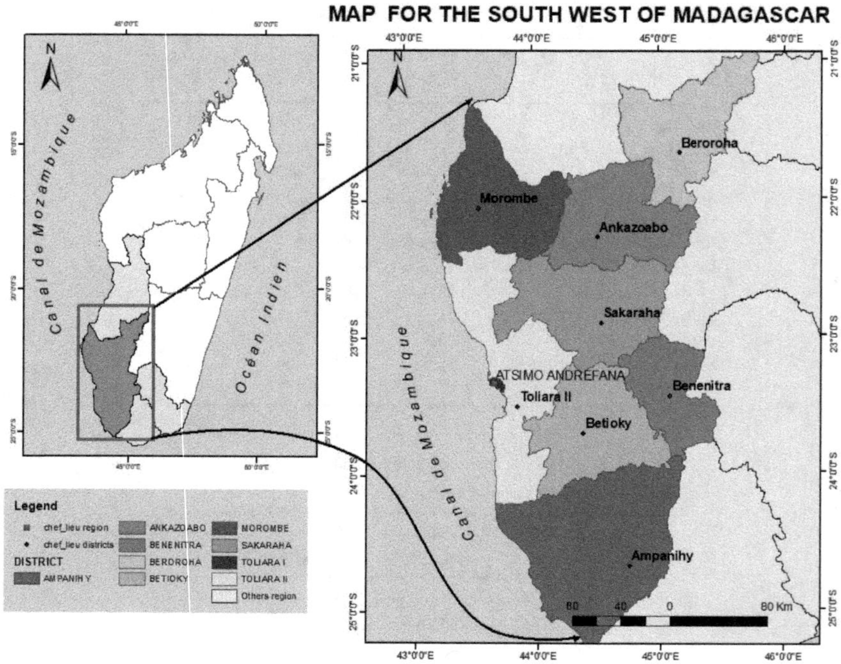

Fig. 1. Area of study.

form: $f(t_1, \ldots, t_p, \beta_0, \ldots, \beta_p) + \varepsilon.\epsilon$ is a quantitative variable with zero mean value. This polynomial representation may be motivated by intuition, the context of the experiment or statistical criteria.

Regression is used to optimize expected results, or to give us an idea of what our curve would look like if there were no residuals in our data.

3.2.2. *Correlation coefficient*

The correlation coefficient is noted ρ in the population.[6]

$$\rho = \frac{\operatorname{cov}(t, y)}{\sqrt{\operatorname{var}(t).\operatorname{var}(y)}}. \tag{4}$$

Its value is between $-1 \leq \rho \leq 1$ and we have the covariance of these variables with the value:

$$\operatorname{cov}(t, y) = \frac{1}{n} \sum_{i=1}^{n} (t_i - \bar{t})(y_i - \bar{y}). \tag{5}$$

3.2.3. *Polynomial regression and correlation coefficient results*

In our work, we chose to use polynomial regression because the cloud of solar irradiance points has the appearance of a curve. Excel was used to calculate the correlation coefficient and plot the variation curve. Fitting the curve of variation gives us the curve in red that we call the polynomial regression curve of equation:

$$y = -18.643t^2 + 455.67t - 2003.3. \tag{6}$$

The regression curve and the value of the correlation coefficient obtained, which is $\rho = 0.98$, clearly show that there is a positive relationship between time and the value of solar irradiance.

3.3. *Interpolation*

Spatial interpolation is the process of using points with known values to estimate values at other unknown points.[5] Spatial interpolation can estimate solar radiation at locations with no recorded data, using known values at nearby locations.

3.4. *Spatial interpolation results*

The Geographic Information System (GIS) using ArcGIS software has enabled us to draw up maps showing the spatial variation of solar radiation in the Atsimo Andrefana region. The method used is spatial interpolation, more precisely Regularized Spline Interpolation with Tension (RST).

We took the mean values for each district and interpolated them to obtain the spatial variation in solar radiation.

Here we have three examples of solar maps showing the spatial variation of solar radiation, chosen to represent the minimum and maximum of the month over the five years studied, with January having the lowest mean value and October the highest. Figure 2 summarizes the histograms illustrating the hourly, monthly, and annual

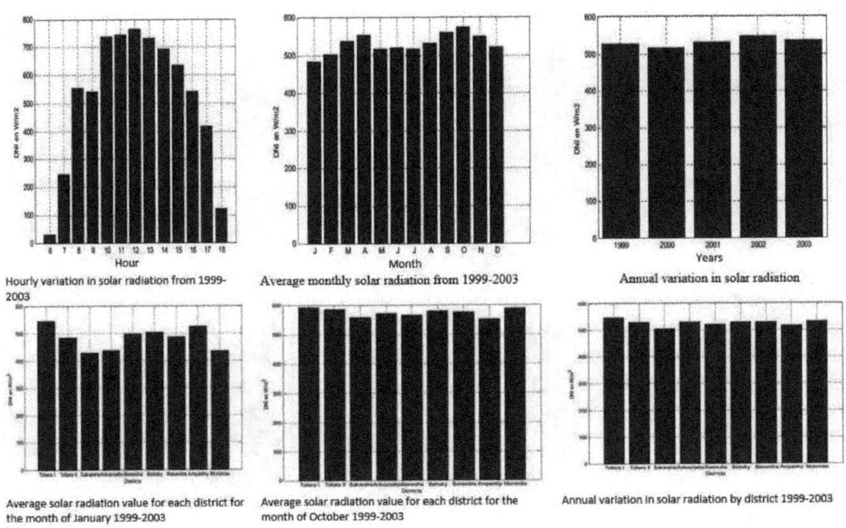

Fig. 2. Average solar radiation values from 1999 to 2003 for Atsimo Andrefana region.

variations of solar radiation, while taking into account the differences observed across the nine districts that make up the region. Figure 3 corresponds to the polynomial regression of solar radiation values as a function of time. The third map illustrates the region's annual spatial variation from 1999 to 2003. Finally, Fig. 4 displays three maps highlighting the spatial variation of solar radiation for the month of October, the month of January, as well as for the entire year over the five-year study period.

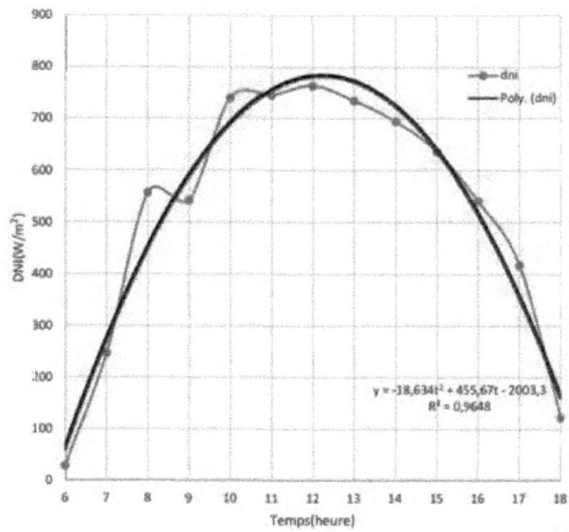

Fig. 3. Polynomial regression curve.

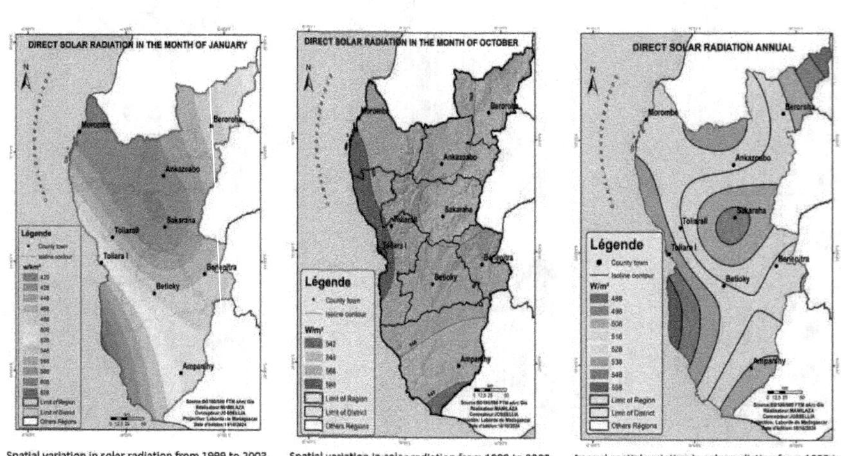

Spatial variation in solar radiation from 1999 to 2003 for the month of january

Spatial variation in solar radiation from 1999 to 2003 for the month of october

Annual spatial variation in solar radiation from 1999 to 2003

Fig. 4. Spatial variation of solar radiation from 1999 to 2003 for Atsimo Andrefana region.

4. Conclusion

In conclusion, we have seen that the Atsimo Andrefana region possesses a gigantic solar deposit that can easily be exploited. To help participants in the energy sector, more specifically those specializing in solar energy, we have drawn up solar maps that clearly explain the variation in sunshine in this region. At the same time, the values obtained from the mean value calculation can help with estimates before making decisions on individual installations. The work we have done is original, but still needs to be perfected. On this point, we'd like to develop interactive maps to give a little more real-time information for a given site. This will make it easier for engineers and researchers to work in the energy sector at a given location.

ORCID

T. Fanjirindratovo https://orcid.org/0009-0003-2131-4447
J. F. Ramaroson https://orcid.org/0009-0001-2429-4125

References

1. K. Bouchouicha, Multispectral modeling of satellite images - application: Quantification of the soil-atmosphere energy balance, doctoral thesis, University of Science and Technology of Oran Mohamed Boudiaf (2017).
2. F. Chabane, Modeling the parameters of thermal conversion of solar energy, Doctorate thesis, University of Mohamed Khider, p. 150 (2015).
3. R. Dambreville, Global solar radiation forecasting by remote sensing for the management of photovoltaic energy production, PhD thesis, University of Grenoble (2015); L. Evain *et al.*, Space and environment workshop, Lyce Roosevelt-Reims (2007).
4. N. Mohamed, Comparative study of solar radiation estimation methods, Magister thesis, University of Ferhat Abbas - Setif (2010).
5. V. Moreau, Statistical estimation of missing data in life cycle inventory, University de Montral (2012).
6. R. Rakotomalala, Correlation analysis/dependence study/variables, University Lumire Lyon (2017).
7. S. H. Rakotoniaina, Method to produce a sunshine map: Case of the south-west region of Madagascar, Electrical Engineering Department, ESPA - Madagascar (2019).

Diffuse fraction for Antananarivo Madagascar

L. Manefemasy* and T. Fanjirindratovo
Physics and Environment Laboratory, University of Toliara,
P. O. Box 185, Toliara 601, Madagascar
**lmanefemasy@gmail.com*

Received 27 February 2025
Accepted 31 July 2025
Published 24 September 2025

The aim of this paper is to find the best model to estimate the hourly solar diffuse fraction by the clearness index for the town of Antananarivo, Madagascar. Per minute data of solar diffuse and global irradiance in horizontal plane during four years were used to build the three-piece defined model. Then, eight models in the literature were applied to compare for the present model by 09 statistical criteria. The result shows that the present model is the best and is the only one that can fulfill all statistical criteria.

Keywords: Solar diffuse fraction; solar global radiation; clearness index; database; linear regression.

1. Introduction

Madagascar, as a tropical country at the Indian Ocean, has a high potential in solar energy, but to be able to use it efficiently, it is important to know the different components of the solar radiation. On a horizontal surface, the solar global radiation is the sum of the direct radiation and the diffuse radiation. There are few weather stations measuring solar radiation in all the countries, and most of them are measuring only the global radiation. On other hand, the cost of purchase, installation and maintenance of these facilities are expensive. Solar radiation as a meteorological parameter has a random and fluctuating nature which makes its study complicated. So, it is important to find a mathematical model to study the different solar parameters. The most available worldwide data are the global radiation in horizontal surface, and it is a principle in this field of research to estimate the other solar parameters from the solar global radiation. For this work, the goal is to estimate the diffuse component from the global radiation, more precisely to express the hourly

*Corresponding author.

solar diffuse fraction by the clearness index for the town of Antananarivo which is the capitol of Madagascar. In this town, there is a weather station that measures among other things, the solar diffuse radiation and the solar global radiation in per minute recording. A database of four years' measurements (2020–2023) from this station is used to fit the model and the result will be compared with the fitting from models available in the literature. The comparison is made by nine statistical criteria. This paper is divided into five sections, Sec. 1 is for the Introduction, after which comes the methodology in Sec. 2 which will expose the following themes: the database, the data quality control, the linear fitting, the statistical criteria and the existing methods in the literature. Section 3 gives the results. Section 4 gives details of the present model, followed by discussion in Sec. 5 and conclusion in Sec. 6.

2. Methodology

2.1. *The data base*

IOS-net, Indian Ocean Network is a project to install a network of stations measuring solar radiation and meteorological parameters within the islands of the Indian Ocean. The project is led by ENERGY-Lab of the University of Reunion which hosts the databases on the site https://galilee.univ-reunion.fr. Then, Madagascar has two stations by this network, the first of which is installed in Antananarivo the capital of the country and the second in Antsiranana. In this work, only the data from Antananarivo will be used because there is too much missing data for Antsiranana for the period 2020–2023. The station measures among other things, the solar diffuse radiation (D) and the solar global radiation (G) with [W/m] as unit and in per minute recording. With time and date and the database, the following solar parameters are calculated by the Spencer algorithm[9]: solar elevation angle (el), solar declination angle (δ), hourly angle (ω), sun-earth distance correction factor (ε) and the extraterrestrial radiation on horizontal plane (G_0).[5] Then, the diffuse and global radiation data between sunrise and sunset time are integrated to get the hourly diffuse irradiation H_d, the hourly global irradiation H and the hourly extraterrestrial global irradiation H_0 on horizontal plane with [Wh/m] as unit. The diffuse fraction k is the ratio of H_d and H. So, for this study the main goal is to find the more precise fitting to get k by k_t.[5–21]

$$k = \frac{H_d}{H}, \tag{1}$$

$$k_t = \frac{H}{H_0}. \tag{2}$$

2.2. *The first set of data quality control*

There are two steps of quality control at this stage. The first is the data quality control from Baseline Station Radiation Network (BSRN) community,[1] there are some intervals for the value of diffuse and global radiation and also for the ratio between them. If the per minute data are outside these intervals, the data are flagged as wrong by a nonzero integer and conversely the flag is set to zero if the data comply with the quality control. Only, data with flag equal to zero will be used. The second test is about the hourly irradiation that should comply with the following tests in Table 1.[2]

2.3. *The envelope quality control*

As mentioned above in the introduction, the global radiation is the sum of the direct radiation and diffuse radiation. The clearness index is an indicator of the cloud coverage, for clear sky, k_t is high and k is low. Conversely, for overcast sky, that is more diffuse, k is high and k_t is low. By Ref. 3, Fig. 1 gives a general representation of the scatter plot of diffuse fraction versus clearness index. From this figure, there is an envelope of normal expectancy where suitable data should be found. The data outside this envelope should be rejected because they can come from misalignment of apparatus or sensors covered by dust or debris (Fig. 2).

Table 1. Diffuse data quality control.

Solar elevation angle	Diffuse fration limit	Clearness index limit	Diffuse limit
$el \leq 7$ (3)	$0 < k_d < 1$ (4)	$0 < k < 1$ (5)	$0 < H_d < 0.8H_0$ (6)

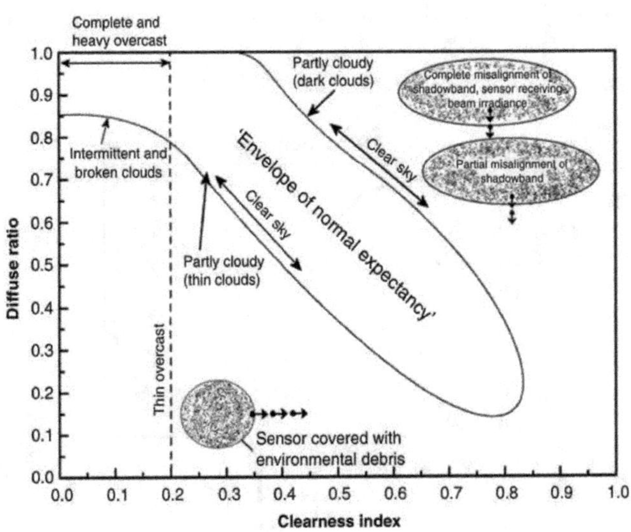

Fig. 1. Envelope of normal expectancy.

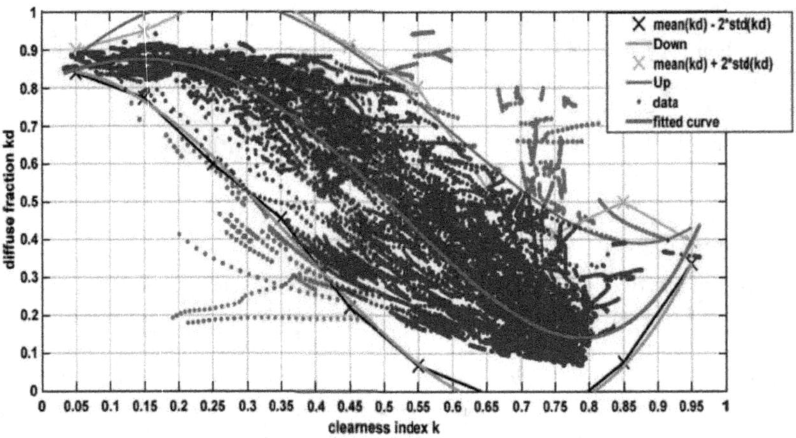

Fig. 2. Scatter plot with envelope quality control.

The following steps should be followed to build this envelope.[3] The interval [0, 1] of k_t is divided into ten equal width bands. For each band, the average value of k, mean(k) and the standard deviation of k, std(k), are calculated. Then, a segmented line envelope is built by an upper segmented line and a lower segmented line. The points corresponding to $\mathrm{mean}(k) + \alpha * \mathrm{std}(k)$ are joined together to build the upper segmented line envelope, and the lower segmented line is built by joining the points corresponding to $\mathrm{mean}(k) - \alpha * \mathrm{std}(k)$. α is a real number between 2 and 3. After what, the two segmented lines are smoothed by polynomial fitting to get the final envelope. Only the data inside the envelope will be used for the final fitting. Figure 3 shows an example from Ref. 4.

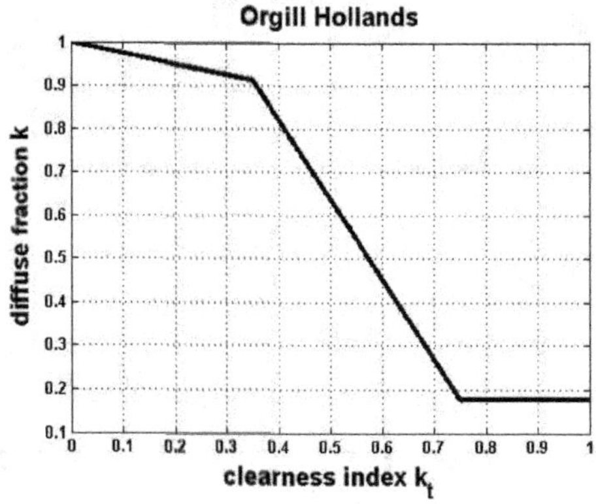

Fig. 3. Orgill and Hollands model.

Here α equals 2. The upper segmented line is in orange, the lower segmented line is in black. Then, the smoothing by polynomial fit gives the final upper part of the envelope, in grey. And the smoothed lower part of the envelope is in green. The useful data are in blue, and rejected data, outside the final envelope, are in red.

2.4. *The fitting of diffuse fraction by clearness index*

According to Refs. 5–21, an example is in Fig. 3 for Orgill and Hollands model. It is called a three-piece wised fitting.The first piece is for very low value of kt, where k is high and almost constant or following a straight line. In the second piece, as k_t increases, k decreases by a polynomial fit. And the third part is for high value of k, where k_t is again constant and minimal. The first and third parts contain a small part of the data, most of it is in the second part.

To estimate the boundaries of the three parts of the fitting, k_t is divided into several bands with 0.025 as width for each band. Then, the average value of k is calculated for each band. The scatter plot of these mean values of k for each band gives an intuition for the limit of the three parts and the fitting for each part.

2.5. *Statistical criteria*

To evaluate the performance, errors, degrees of precision and significance at a defined threshold of the model, the following criteria calculating the difference of the measured value of k and calculated value of k by the model are proposed in Table 2. The Mean Bias Error (MBE) and Normalized Mean Bias Error (NMBE%) measure the behavior of the model for the long-term. On the other hand, Mean Absolute Bias Error (MABE), Root Mean Square Error (RMSE) and their respective normalized value NMABE% and NRMSE%, estimate the short-term comparison of the model.

Table 2. Statistic criteria efficiency.

Statistical parameters	Ideal value for the best model
$\mathrm{NSE} = 1 - \frac{\sum_{i=1}^{n_o} (k_{\mathrm{mes}_i} - k_{\mathrm{cal}_i})^2}{\sum_{i=1}^{n_o} (k_{\mathrm{kal}_i} - \mathrm{mean}(k_{\mathrm{mes}_i}))^2}$	1
$R = \frac{\sum_{i=1}^{n_o} ((k_{\mathrm{cal}_i} - \mathrm{mean}(k_{\mathrm{cal}_i})) * (k_{\mathrm{mes}_i} - k_{\mathrm{mes}_i}))}{\sqrt{\sum_{i=1}^{n_o} (k_{\mathrm{cal}_i} - k_{\mathrm{cal}_i})^2 * (k_{\mathrm{mes}_i} - k_{\mathrm{mes}_i}))^2}}$	1
$R^2 = (R)^2$	1
$\mathrm{MBE} = \frac{1}{n_o} * \sum_{i=1}^{n_o} (k_{\mathrm{cal}_i} - k_{\mathrm{mes}_i})$	0
$\mathrm{NMBE} = \frac{\mathrm{MBE}}{\frac{1}{n_o} * \sum_{i=1}^{n_o} k_{\mathrm{mes}_i}} * 100$	0
$\mathrm{MABE} = \frac{1}{n_o} * \sum_{i=1}^{n_o} k_{\mathrm{cal}_i} - k_{\mathrm{mes}_i}$	0
$\mathrm{NMABE} = 1 - \frac{\sum_{i=1}^{n_o} (k_{\mathrm{mes}_i} - k_{cal_i})^2}{\sum_{i=1}^{n_o} (k_{\mathrm{kal}_i} - \mathrm{mean}(k_{\mathrm{mes}_i}))^2}$	0
$\mathrm{RMSE} = \sqrt{\frac{1}{n_o} * \sum_{i=1}^{n_o} (k_{cal_i} - k_{mes_i})^2}$	0
$\mathrm{NRMSE} = \frac{\mathrm{RMSE}}{\frac{1}{n_o} \times \sum_{i=1}^{n_o} k_{\mathrm{mes}_i}} * 100$	0
$t_{\mathrm{stat}} = \sqrt{\frac{(n_o - 1)\mathrm{MBE}^2}{\mathrm{RMSE}^2 - \mathrm{MBE}^2}}$	$\lvert t_{\mathrm{stat}} \rvert < t_{\mathrm{crit}}$

Table 3. Some models for estimating the diffuse fraction of solar radiation from global solar radiation.

Models	Locations	Date	Diffuse fraction	Constraints
Orgil and	Toronto	1967–	$k = 1 - 0.249k_t$	$0.00 < k_t \leq 0.35$
Holand[5]	Canada	1971	$k = 1.157 - 1.84k_t$	$0.35 < k_t \leq 0.75$
			$k = 0.177$	$0.75 < k_t < 1$
Erbs	Highett,	1966–	$k = 1 - 0.99k_t$	$0.00 < k_t \leq 0.22$
et al.[6]	Australia	1969	$k = 0.9511 - 0.1604k_t + 4.388k_t^2 - 16.638k_t^3 + 12.336k_t^4$	$0.22 < k_t \leq 0.80$
	(38S)		$k = 0.165$	$0.80 < k_t \leq 1$
Reindl	European,	1979–	$k = 1.02 - 0.248k_t$	$0.00 < k_t < 0.30$
et al.[7]	North America	1982	$k = 1.45 - 1.67k_t$	$0.30 \leq k_t < 0.78$
	(28.4N-59.56N)		$k = 0.147k_t$	$0.80 \leq k_t < 1$
Hawlader[8]	Paya lebar,	1962	$k = 0.915$	$0.000 < k_t \leq 0.225$
	Singapore		$k = 1.1389 - 0.9422k_t - 0.3878k_t^2$	$0.225 < k_t < 0.775$
	(1"22'N)		$k = 0.215$	$0.775 \leq k_t < 1$
Spencer[9]	Australia		$k = a_3 - b_3k_t - 0.3878k_t^2$	$0.35 < k_t < 0.75$
			$a_3 = 0.94 + 0.0118\lvert\varphi\rvert, b_3 = 1.185 + 0.0135\lvert\varphi\rvert$	
	Madras, Inde	1983–	$k = 1.0086 - 0.178k_t$	$0.00 < k_t \leq 0.24$
	(13.00N)	1987	$k = 0.9686 - 0.13250k_t + 1.4183k_t^2 - 10.1862k_t^3 + 8.3733k_t^4$	$0.24 < k_t \leq 0.80$
Saioa *et al.*[10]			$k = 0.197$	$0.80 < k_t \leq 1$
Boland	Victoria		$k = \frac{1}{1+exp(7.997(k_t-0.586)}$	
et al.[11]	Australia 38.09S			
Boland	Adelaide,		$k = \frac{1}{1+exp(-5.0033+8.6025k_t)}$	
et al.[11]	Australia 34.9S			
Saioa	Aire,	1978–	$k = 0.995 - 0.081k_t$	$0.00 < k_t \leq 0.21$
et al.[10]	Mediterranean	1996	$k = 0.724 + 2.738k_t - 8.32k_t^2 + 4.967k_t^3$	$0.21 < k_t \leq 0.76$
	(37.02-44.08)		$k = 0.18$	$0.76 < k_t \leq 1$
Saioa	Sao Paulo,	1994–	$k = 1$	$0.00 < k_t \leq 0.17$
et al.[10]	Brasilia	1999	$k = 0.97 + 0.8k_t - 3k_t^2 - 3.11k_t^3 + 5.2k_t^4$	$0.17 < k_t \leq 0.75$
	(2333'35"S)		$k = 0.18$	$0.75 < k_t \leq 1$
Karatasou	Athens, Greece	1996–	$k = 0.9995 - 0.05k_t - 2.4156k_t^2 + 1.4926k_t^3$	$0.00 < k_t \leq 0.78$
et al.[12]	(38.049N)	1998	$k = 0.20$	$0.78 < k_t \leq 1$

Table 3. (*Continued*)

Models	Locations	Date	Diffuse fraction	Constraints
Soares	Sao Paulo,	1998–	$k = 1$	$0.00 < k_t \leq 0.17$
et al.[13]	Brasilia	2001	$k = 0.90 + 1.1k_t - 4.5k_t^2 + 0.01k_t^3 + 3.14k_t^4$	$0.17 < k_t \leq 0.75$
	(2333'35"S)		$k = 0.215k_t$	$0.75 < k_t \leq 1$
Jacovides	Athalasse,	1998–	$k = 0.987$	$0.0 < k_t \leq 0.1$
et al.[14]	Cyprus	2002	$k = 0.94 + 0.937k_t - 5.01k_t^2 + 3.32k_t^3$	$0.1 < k_t \leq 0.8$
	(35N)		$k = 0.177$	$0.8 < k_t \leq 1$
Lam and	Hong Kong,	1991–	$k = 0.997$	$0.00 < k_t \leq 0.15$
Li[15]		1994	$k = 1.237 - 1.361k_t$	$0.15 < k_t \leq 0.70$
			$k = 0.273$	$0.70 < k_t \leq 1$
Muneer and Saluja[3]	New Delhi,		$k = 0.95$	$0.000 < k_t \leq 0.175$
	Inde		$k = 0.9698 + 0.4353k_t - 3.4499k_t^2 + 2.1888k_t^3$	$0.155 < k_t \leq 0.755$
			$k = 0.273$	$0.775 < k_t \leq 1$
Fanjirindratovo,	Island, Indian	2007–	$k = 0.8161 + 0.7096k_t - 1.5255k_t^2 - 5.074k_t^3 + 5.6980k_t^4$	
	Ocean[4]	2020		
Bugler[16]	Highett, Australia	1966–	$k = 0.94$	$0.0 < k_t \leq 0.4$
	(38S)	1970	$k = \frac{1.29 - 1.19k_t}{1 - 0.334k_t}$	$0.4 < k_t < 1$
Al Riahi,	Fudhaliyah,	1984–	$k = 0.932$	$0.00 < k_t \leq 0.25$
et al.[17]	Iraq	1987	$k = 1.293 - 1.631k_t$	$0.25 < k_t \leq 0.70$
	(3314'N)		$k = 0.151$	$0.70 < k_t \leq 1$
Chick	Algerie	1990–	$k = 1 - 0.232k_t$	$0.000 < k_t \leq 0.175$
et al.[18]	(22.8N-36.43N)	1992	$k = 1.293 - 1.631k_t$	$0.175 < k_t < 0.870$
	(3314'N)		$k = 0.203$	$0.870 \leq k_t < 1$
Elminir	Egypt	1999–	$k = 0.8526 - 1.1778k_t$	$0.00 < k_t \leq 0.22$
et al.[19]	(2358'-3005'N)	2002	$k = 0.814 - 1.106k_t + 0.366k_t^2 - 0.997k_t^3 + 1.221k_t^4$	$0.22 < k_t \leq 0.80$
			$k = 0.213$	$0.80 < k_t \leq 1$

Table 3. (*Continued*)

Models	Locations	Date	Diffuse fraction	Constraints
Ruiz-Arias *et al.*[20]	Albacete, Spain (39.00N)	2002–2006	$k = 0.086 + 0.88exp(-exp(-3.877 + 6.138k_t))$	
Macagnan *et al.*[21]	Madrid, Spain (39.00N)	1975–1991	$k = 0.996 + 0.424k_t - 0.586k_t^2$	$0.00 < k_t \leq 0.22$
			$k = 1.11 - 0.203k_t - 2.52k_t^2 + 0.617k_t^3 + 1.063k_t^4$	$0.22 \leq k_t < 0.70$
			$k = -0.0169 - 0.99k_t + 1.63k_t^2$	$0.70 < k_t \leq 1$
Elminir *et al.*[19]	Egypt (2358'-3005'N)	1999–2002	$k = 0.8526 - 1.1778k_t$	$0.00 < k_t \leq 0.22$
			$k = 0.814 - 1.106k_t + 0.366k_t^2 - 0.997k_t^3 + 1.221k_t^4$	$0.22 < k_t \leq 0.80$
			$k = 0.213$	$0.80 < k_t \leq 1$
Ruiz-Arias *et al.*[20]	Albacete, Spain (39.00N)	2002–2006	$k = 0.086 + 0.88exp(-exp(-3.877 + 6.138k_t))$	
Macagnan *et al.*[21]	Madrid, Spain (39.00N)	1975–1991	$k = 0.996 + 0.424k_t - 0.586k_t^2$	$0.00 < k_t \leq 0.22$
			$k = 1.11 - 0.203k_t - 2.52k_t^2 + 0.617k_t^3 + 1.063k_t^4$	$0.22 \leq k_t < 0.70$
			$k = -0.0169 - 0.99k_t + 1.63k_t^2$	$0.70 < k_t \leq 1$

The Nash–Sutcliffe Equation (NSE), correlation coefficient R and R^2 and t_{stat} estimate the quality of the fitting and especially for t_{stat} the validity of the model. By Student's table, according to the number of the data, there is a limit value for t_{stat}. If the calculated t_{stat} for the model is lower than this limit, the model is validated, conversely if t_{stat} is higher than this limit, the model is not validated. Table 2 shows also the ideal value for each criterion for the best model.

2.6. *The existing models*

Table 3 gives a list of 22 existing models in the literature, 15 of them are three-piecewise. Five are one-piecewise, and 2 are two-piecewise.

3. Results

3.1. *Result of the data quality control*

Table 4 gives the number of initial data, the number of data removed by the control quality, and at the end the number of useful data and Fig. 4 shows the scatter plot with the smooth envelope, the useful data are in blue and the rejected data in red. The value of α is 2.3.

Table 4. Result of the data quality control.

Initials data	18 873
Flag (BSRN test)	460
$h < 7°$ (test 1)	1309
$k_t > 1, k > 1$ (test 2)	0
Envelop (test 3)	329
Valid data	16 775

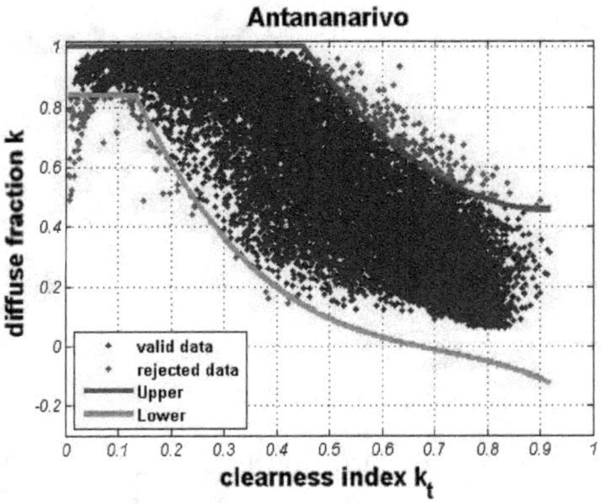

Fig. 4. Rejected and valid data by envelope quality control.

Upper for $0 < k_t \leq 0.45$ $k = 1$ for $0.45 < k_t \leq 1$ $k = 2.553 - 4.5749k_t + 2.4971k_t^2$
Lower for $0 < k_t \leq 0.135$ $k = 0.840$ for $0.135 < k_t \leq 1$ $k = 1.4480 - 5.4648k_t + 7.2112k_t^2 - 3.4026k_t^3$.

4. The Present Model

The first action was to calculate the average value of k for each band of$k_t$with 0.025 of width. Then, the scatter plot of these data gives intuition for the model. Figure 5 shows the plot (in blue marker) of the mean value of k by 0.025 bandwidth of k_t. Then, the model is plotted in black line. For very low values of k_t, the value of the diffuse ratio approaches unity, overcast sky region. We note that in the range: $0 < k_t < 0.176$ the days are very cloudy and represent 6.31 % of the data over a 4-year period, and we suggest a constant value of $k = 0.945$.
– Beyond this region, the diffuse ratio decreases significantly until reaching its minimum, a region considered to be partially overcast. 90.07 % of the data occupy this region and lie in the interval $0.176 \leq k_t \leq 0.801$ and we can recommend a polynomial function of degrees 2 by: $k = 1.09 - 0.78k_t - 0.48k_t^2$.
– Finally, after the descent of the function, it stabilizes relatively for k_t close to 1 (clear sky region), represented by 3.62 % of the data and also in our case based on

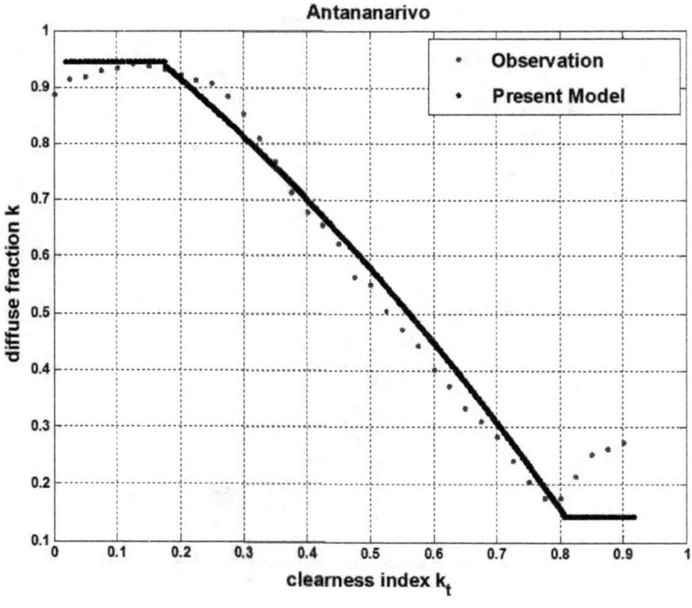

Fig. 5. Scatter plot of mean value of k for each band of 0.025 width of k_t, and the present model fitting.

$0.801 < k_t \leq 1$ and correspond to days well released, which has an estimated value around $k = 0.145$. The model is then defined by the following function:

$$\begin{aligned} k &= 0.945 && \text{for } 0.000 < k_t < 0.176, \\ k &= 1.09 - 0.78k_t - 0.48k_t^2 && \text{for } 0.176 \leq k_t \leq 0.806, \\ k &= 0.145 && \text{for } 0.806 < k_t < 1. \end{aligned}$$

4.1. *Statistical test results*

Table 5 gives the result of the statistical criteria for the existing models and the model in this work. On the last row is shown the relative difference between the present model and the best model for each criterion. For NSE, R^2 and NRMSE%, the best model is Jacovides, but the present model is very close because the difference is only from −0.9% to 0.5%. For NMABE%, the best model is Reindl, and the relative difference with the present model is 5%. For NMBE% and t_{stat}, the present model is by far the best, because the relative difference with the model in second position, which is Hawlader, is more than 3000%. But the important thing is for the t_{stat}, with more than 16000 data points, the limit of the t_{stat} is 1.96. And by Table 5, all the previous models have a t_{stat} higher than 1.96, whereas it is 0.105 for the present

Table 5. Results of statistical criteria test.

Models	n_o	NSE	R	R^2	NMBE%	NMABE%	NRMSE%	t_{stat}
Orgill Holland	16 775	0.758	0.888	0.788	6.46	18.84	26.70	32.328
Erbs al	16 775	0.737	0.887	0.787	7.35	19.53	27.84	35.466
Reindl al	16 775	0.774	0.894	0.799	5.30	18.62	25.81	27.198
Hawlader	16 775	0.798	0.894	0.8	−0.78	19.41	24.39	4.148
Spencer	16 775	0.01	0.888	0.789	−42.90	46.76	53.96	169.721
Chandrasekaran Kumar	16 775	0.763	0.891	0.794	7.56	19.14	26.41	38.71
Boland al	16 775	0.745	0.886	0.784	6.762	19.24	27.38	33.003
Miguel al	16 775	0.764	0.892	0.795	6.506	18.92	26.34	33.009
Oliveira al	16 775	0.782	0.89	0.793	1.305	19.43	25.34	6.68
Karatasou al	16 775	0.795	0.893	0.798	−1.51	19.79	24.54	7.997
Soares al	16 775	0.779	0.891	0.794	−4.70	19.69	25.51	24.317
Jacovides	16 775	0.799	0.895	0.8	1.10	18.84	24.34	5.884
Lam et Li	16 775	0.785	0.888	0.788	2.12	20.56	25.16	10.966
Muneer	16 775	0.792	0.892	0.797	3.69	19.25	24.76	19.544
Fanjiry	16 775	0.751	0.893	0.797	−11.55	21.58	27.09	61.065
Bugler	16 775	0.127	0.855	0.731	39.79	41.02	50.6	164.12
Al Riahi al	16 775	0.728	0.889	0.791	−12.65	21.76	28.28	64.777
Chikh al	16 775	0.792	0.891	0.794	−0.89	20.02	24.74	4.694
Elminir al	16 775	0.736	0.887	0.788	7.34	19.55	27.84	35.426
Ruiz-Arias al	16 775	0.773	0.883	0.78	3.10	19.24	25.82	15.681
Macagnan al	16 775	0.753	0.882	0.777	−5.46	21.14	26.97	26.776
Present	16 775	0.795	0.892	0.796	−0.019	19.50	24.556	0.105
Difference in % between the best model and this one		0.5%	0.3%	0.5%	0%	−5%	−0.9%	%

model. So, all models in the literature are not validated for the town of Antananarivo, and only the present model is suitable for the present site.

5. Discussion

5.1. *About the quality control*

There are 2098 rejected data among the 18 873 initial data, that represents about 11%. It shows the importance of the quality control at the per minute raw data, and also for the hourly irradiation data. For the envelope quality control, the value of α is 2.3, the reason of this choice is that one should find the suitable coefficient that ensures the final result fulfills all the statistical criteria, especially the t_{stat}. It was found that the value of 2.3 of α is the appropriate value to satisfy all statistical conditions.

5.2. *About the fitting of the present model*

Table 3 shows that most of the fitting of the existing models is three-piecewise but some or few of them are two-piecewise. For the present model, the condition to choose a fitting model is always to fulfill all statistical criteria especially the t_{stat}. The result shows that the present model is the best. Some of the existing models give also a good result about the long-term and short-term behaviors, and also for NSE and R^2 test, but all of them fail on the t_{stat} test and condition. The present model, not only satisfies the t_{stat} condition, but also gives a good result for the long-term or short-term comparison. If the present model is the best for a criterion, like NMBE%, the relative difference with the second-best method, in the list, is very high. And conversely, if another method is the best for a criterion, the present method is very close as 0.3% to 5% as a relative difference. It means that for long-term comparison the present method is, by far, the best and for short-term comparison, it is almost the same for present and previous methods. But in t_{stat} condition, only the present method is appropriate for the town of Antananarivo.

6. Conclusion

The present method is the only one validated to estimate the diffuse fraction from the clearness index for the town of Antananarivo. Of course, the other methods were fitted by data from another place in the world, and it is the reason why they are not able to fulfill the t_{stat} criterion, even if they give some good results for other criteria. Geographically, Madagascar is a member of the Indian Ocean community. So, the next step is to extend the study not only to other towns in Madagascar, but also for other islands in the Indian Ocean, to see the applicability of the present model.

ORCID

L. Manefemasy https://orcid.org/0009-0005-2677-1537
T. Fanjirindratovo https://orcid.org/0009-0003-2131-4447

References

1. C. N. Long and E. G. Dutton, BSRN global network recommended QC tests, V2.0. BSRN Tech. Rep. (2022), https://epic.awi.de/30083/1/BSRN/recommended QC tests
2. S. Younes, R. Claywell and T. Muneer, *Energy* **30**, 1533 (2005).
3. T. Muneer and G. S. Saluja, *Energy Convers. Manage.* **25**, 443 (1985).
4. T. Fanjirindratovo, Hourly diffuse fraction correlation, in *Proc. High-Energy Physics Int. Conf.* Antananarivo, Madagascar, 2022, pp. 10–16.
5. J. F. Orgill and K. G. Hollands, *Sol. Energy* **19**, 357 (1976).
6. D. G. Erbs, S. A. Klein and J. A. Duffie, *Sol. Energy* **28**, 293 (1982).
7. D. T. Reindl, W. A. Beckman and J. A. Duffie, *Sol. Energy* **45**, 1 (1990).
8. M. N. A. Hawlader, *Int. J. Ambient Energy* **5**, 31 (1984).
9. J. W. Spencer, *Sol. Energy* **29**, 19 (1982).
10. E. B. Saioa, J. G. Eulalia and T. Muneer, *Energies* **13**, 2906 (2020).
11. J. Boland, H. Jing and R. Barbara, *Renew. Sustain. Energy Rev.* **28**, 749 (2013).
12. S. Karatasou, M. Santamouris and V. Geros, *Int. J. Sustain. Energy* **23**, 1 (2003).
13. J. Soares, A. P. Oliveira, Z. B. Marija, M. Primoz, F. E. Joao and J. M. Antonio, *Appl. Energy* **79**, 201 (2004).
14. C. P. Jacovides, F. S. Tymvios, V. D. Assimakopoulos and N. A. Kaltsounides, *Renew. Energy* **31**, 2492 (2006).
15. J. C. Lam and D. H. W. Li, *Build. Environ.* **31**, 527 (1996).
16. J. W. Bugler, *Sol. Energy* **19**, 477 (1977).
17. M. Al-Riahi, N. Al-Hamdani and K. Tahir, *Renew. Energy* **2**, 451 (1992).
18. M. Chikh, A. Mahrane and M. Haddadi, *Energy Procedia* **18**, 1068 (2012).
19. H. K. Elminir, Y. A. Azzam and F. I. Younes, *Energy* **32**, 1513 (2007).
20. J. A. Ruiz-Arias, H. Alsamamra, J. Tovar-Pescador and D. Pozo-Vazquez, *Energy Convers. Manage.* **51**, 881 (2010).
21. M. H. Macagnan, E. Lorenzo and C. Jimenez, *Int. J. Sol. Energy* **16**, 1 (1994).

Estimating sunshine duration from global solar radiation for Antananarivo*

T. Fanjirindratovo † and O. Ramiarinjanahary
Physics and Environment Laboratory,
University of Toliara, P. O. Box 185, Toliara 601, Madagascar
†fanjiry@gmail.com

D. Calogine and O. Chau
PIMENT Laboratory, University of Reunion Island 97400 Saint-Denis,
La Réunion, France

M. H. Andrianantenaina
Laboratory of Mechanics Energetic and Environment,
University of Fianarantsoa, Andrainjato, Fianarantsoa 301, Madagascar

Received 24 February 2025
Revised 2 August 2025
Accepted 4 August 2025
Published 8 September 2025

The aim of this paper is to find the best algorithm, among the existing, to estimate sunshine duration from global solar irradiation for the town of Antananarivo-Madagascar. Five methods in the literature based on Pyranometric Methods have been proposed for four years data as the Step Algorithm, the Carpentras or Meteo France Algorithm, the Slob & Monna Algorithm, the revised Slob & Monna Algorithm and the Linear Algorithm. Methods comparison showed that the best method is the Meteo France Algorithm for Antananarivo and the suitable coefficients are $A = 0.65$ and $B = 0.01$. Sunshine duration from solar global radiation results has been discussed. Data analysis has been performed at the parameters statistical significance and in residual analysis for the model fit quality. Results showed that the mean daily differences between the reference and the Meteo France algorithm in hour are 0.0006, the standard deviation of the daily differences in hour is 0.55, the four years total cumulative differences in hour are 0.83 and the Normalized Root Mean Square Error is 8.54%. This study is a part of a global project to study the sunshine duration from global irradiation for the islands in Indian Ocean.

Keywords: Sunshine duration; global radiation.

*Talk given at the 14th International Conference in High-Energy Physics — HEPMAD24, 21–26th October 2024, Antananarivo Madagascar.

†Corresponding author.

1. Introduction

According to the World Meteorological Organization (WMO) in its Guide to Instruments and Methods of Observation in Ref. 1, the sunshine duration is the sum of the sub-period for which the direct solar beam irradiance exceeds 120 $\mathrm{Wm^{-2}}$. The first way to measure it is to use a pyrheliometer installed in a two-axis solar tracker. The second and indirect way measures first the global and diffuse irradiation, then the difference between them, which gives the direct irradiance on horizontal surface, is divided by the sine of solar angle elevation to get the solar direct beam irradiance. Having observed that the most available data in solar radiation are global radiation on horizontal surface, it is common in studies of solar radiation to find methods for estimating other parameters from the global radiation. For sunshine duration, these methods are called pyranometric methods and according to Refs. 2–4, they are based on a long-time comparison between the direct solar beam irradiance measured by a pyrheliometer and the global irradiance by a pyranometer. According to Ref. 5, abundant documentation is not available relating to the applicability of these methods for site in the tropics inside the southern hemisphere. It is for this reason that the authors decided to conduct a global project of estimating the sunshine duration by the global irradiation on horizontal surface for islands in Indian Ocean that regroups Madagascar, Reunion Island, Mauritius Island, Comoros, Seychelles Island, and Maldives Island. The study started from Reunion Island, in the locality of Saint Pierre, whose report can be found in Ref. 5, and this paper, in the following, will focused on the town of Antananarivo, Madagascar. There are mainly five algorithms in the literature that include the Step Algorithm [4,6], the Meteo France or Carpentras Algorithm,[3,4,6,9] the Slob & Monna Algorithm in its two versions[2,7,8] and the Linear Algorithm.[2,7] According to Refs. 4–6 the best method is the Meteo France Algorithm. But for whichever site, one should find the appropriate value for the main coefficients of this algorithm. For this paper, the five algorithms will be applied, and the best coefficients from the previous study for Reunion Island will be used for the Meteo France Algorithm. Afterward, the Meteo France Algorithm will be improved to well fit the present site. This paper is divided into five sections: Section 1 is the introduction, then the methodology will be presented in Sec. 2, Sec. 3 is for the results, Sec. 4 about the improvement of Meteo France Algorithm, and finally, Sec. 5 is the discussion and conclusion.

2. Materials and Methods

2.1. *The data base*

The data came from ENERGY-Lab (2020-2023).[10] Solar and meteorological data were collected from Antananarivo station (Madagascar) by the ENERGY-Lab at the University of La Reunion over the period of 2020–2023. The Antananarivo station, latitude = −18.8983 and longitude = 47.5464, is a member of the IOS-net Indian Ocean Solar Network that has one or more similar stations in each country of Indian

Ocean. It is a per minute recording of solar and different meteorological parameters, especially the global radiation in horizontal plane (G) and diffuse radiation in horizontal plane (D). The other solar parameters like declination angle, solar azimuth angle (az), solar elevation angle (el), sun-earth distance correction factor (ε) are calculated according to the Michalskys Algorithm in Ref. 11. The value of the mean extra-terrestrial global irradiance I_0 is according to Ref. 12.

2.2. *The reference sunshine duration*

As it is a main and fundamental parameter, the sine of the solar elevation angle el notes μ_0 as

$$\mu_0 = \sin(el). \tag{1}$$

The direct solar beam irradiance I is

$$I = \frac{G - D}{\mu_0}. \tag{2}$$

The reference sunshine duration (SDref) is estimated per minute.

If $I \geq 120\,\mathrm{Wm}^{-2}$ and $el \geq 3°$, then SDref = 1 min, otherwise SDref = 0.

The limit of 3° is to avoid abnormal high value of I for low elevation angle during sunrise and sunset.[2–4]

2.3. *Data quality control*

The Baseline Station Radiation Network (BSRN) community has developed a quality control protocol for solar radiation data.[13] There are different defined intervals for the value of each parameter and also for their ratio between them. A flag is associated with each per minute measurement. If the measured data are inside the intervals, the flag is set to zero and the data are classified as good. Otherwise, if the value is outside the intervals, it is considered as erroneous, so the flag is set to a nonzero number. Only data with flag equals to zero will be used in the study.

2.4. *The pyranometric methods*

There are five algorithms in the literature. Because lack of room, only the Meteo France Algorithm will be detailed here. The reader is requested to refer to the appropriate references for other algorithms. Mainly, these algorithms compare the averaged value of the global radiation during a sub-period with a predefined threshold. If the global radiation is higher or equal than this limit, there is sunshine during this sub-period, otherwise there is no sunshine.

2.4.1. *Carpentras or meteo france algorithm (MFA)*

The details are in [3,4,6,9]. The per minute value of the global radiation is compared to a threshold. SDMFA is the sunshine duration for this method.

Table 1. A and B coefficients for meteo france algorithm.

BSRN station	Latitude	A	B
Momote	−2	0.68	−0.06
Tamanrasset	22	0.57	0
Carpentras	44	0.51	0.05
Cabauw	52	0.57	0.06
Toravere	58	0.54	0.06
Saint Pierre	−21.34	0.64	−0.09
Saint Pierre	−21.34	0.63	−0.04
Saint Pierre	−21.34	0.63	−0.05

If $G \geq G\text{th}$, and $el \geq 3°$, then SDMFA = 1 min, otherwise SDSA = 0

$$G_{\text{th}} = 1080\left(A + B\cos\left(\frac{360d}{365}\right)\right)\mu_0^{1.25}, \tag{3}$$

d is the day number as $d = 32$ for February, 1st. The key parameter of the algorithm is the value for A and B coefficients. Table 1 lists some value for these coefficients, the first five rows are for some BSRN stations. The last three rows are the best coefficients from the previous study in Reunion Island.[5]

Table 1 shows that most of the coefficients were fitted for country in northern hemisphere, unless for Momote station. As reported in Ref. 5, the study for Reunion Island gives the last three couples of coefficients. The best couple of coefficients is the one that gives the lowest difference between the algorithm and the reference, these three couples give this minimum with a minor difference between them for Reunion Island.

2.5. *Statistical parameters*

The statistical parameters are to measure the performance of each algorithm, which is based on the difference between the sunshine duration from the algorithm and the sunshine duration from the reference. The main data are the daily sunshine duration (SD) in hour per day. If SDref is the daily reference sunshine duration, SDx the daily sunshine duration for any algorithm, and N the number of experimental points, the statistical tests are as shown in Table 2.

The totdif and rtotdif estimate the long-term behavior of the model. On the other hand, RMSE and NRMSE (%) estimate the models behavior for the short term. The curve of the cumulative difference is also an important tool to evaluate the algorithms performance, the plot of this curve should be closest to the zero axis (SD = 0) for the best algorithm.

2.6. *The methodology of the present study*

The five methods will be applied for the per minute four years data between sunrise and sunset. For the Meteo France Algorithm, there will be four couples of A and B coefficients value as shown in Table 3.

Table 2. Statistical parameters.

Notation	Mathematical expression	Meaning
totdif	$\text{totdif} = \sum_{i=1}^{i=N}(\text{SDx}_i - \text{SDref}_i)$	Total cumulative difference
rtotdif	$\text{rtotdif}[\%] = 100\frac{\text{totdif}}{\sum_{i=1}^{i=N}\text{SDref}_i}$	Total relative cumulative difference
RMSE	$\text{RMSE} = \sqrt{\frac{1}{N}\sum_{i=1}^{i=N}(\text{SDx}_i - \text{SDref}_i)^2}$	Root Mean Square Error
NRMSE (%)	$\text{NRMSE}[\%] = 100\frac{\text{RMSE}}{\frac{\sum_{i=1}^{i=N}\text{SDref}_i}{N}}$	Normalized Root Mean Square Error
mdd	$mdd = \frac{1}{N}\sum_{i=1}^{i=N}(\text{SDx}_i - \text{SDref}_i)$	Mean Value of the daily differences between reference and algorithm
sdd	$sdd = \sqrt{\frac{1}{N-1}\sum_{i=1}^{i=N}(\text{SDx}_i - \text{SDref}_i - mdd)^2}$	Standard deviation of the difference between algorithm and the reference

Table 3. Coefficients for the meteo france algorithm.

Site	Latitude	*A*	*B*
Momote	−2	0.68	−0.06
Saint Pierre coef 1	−21.34	0.64	−0.09
Saint Pierre coef 2	−21.34	0.63	−0.04
Saint Pierre coef 3	−21.34	0.63	−0.05

3. Results

3.1. *For the total and relative cumulative differences between algorithms and the reference*

By Fig. 1, the algorithm that has the best value, nearest to zero, is the Meteo France Algorithm with Momotes station coefficients. The total cumulative difference is −218 hours that is −2.37% in relative value. The second (2.42%), third (2.63%) and fourth positions (2.59%) are still the Meteo France Algorithm but with coefficients from Reunion Island.

Figure 2 shows the result for the NRMSE%. The best value is for the Meteo France Algorithm with Momotes station coefficients with 8.63% as NRMSE%. The second position is the Meteo France Algorithm with the second set of coefficients from Saint Pierre locality (10% as NRMSE%), and third position is the Linear Algorithm (10.15% as NRMSE%). In the fourth position is the Meteo France Algorithm with the third set of coefficients from Saint Pierre locality (10.38% as NRMSE%). As it can be seen in Fig. 3, for the mean of the daily differences, the best value is again for Meteo France Algorithm with coefficients from Momotes station as −0.15 h for value. The second (0.16 h), third (0.17 h) and fourth positions (0.18 h) are still the Meteo France Algorithm with coefficients from the Reunion Islands study. For the standard deviation, the best value is for the Linear Algorithm (0.51 h)

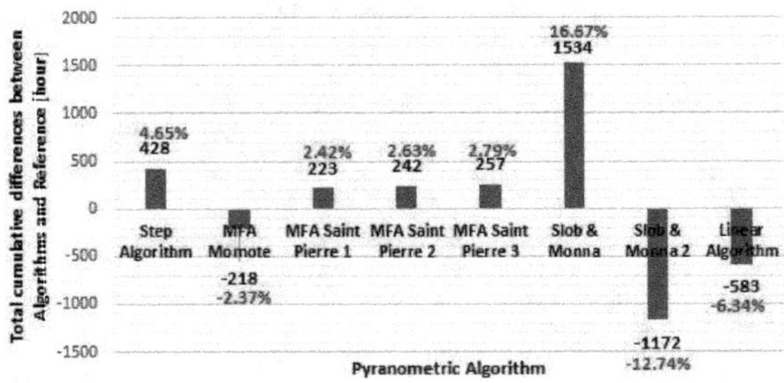

Fig. 1. Total and relative cumulative differences between each algorithm and the reference.

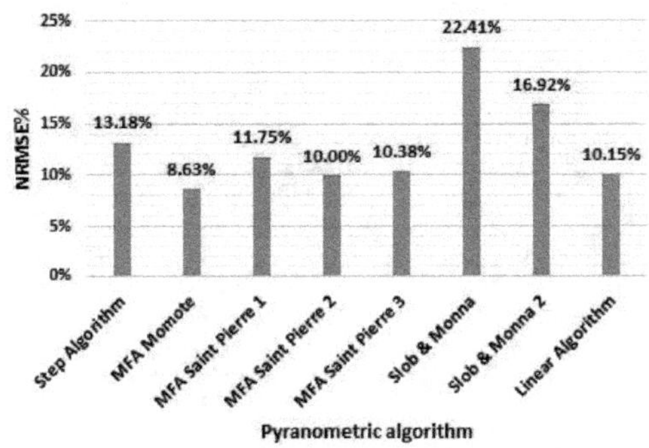

Fig. 2. The Normalized Root Mean Square Error for each algorithm.

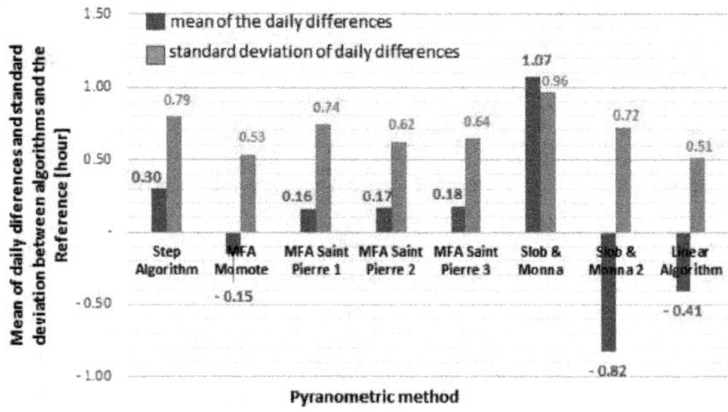

Fig. 3. The mean and standard deviation of the daily differences between algorithms and the reference.

followed immediately by the Meteo France Algorithm with Momotes station coefficients (0.53 h) and in third and fourth positions is the Meteo France Algorithm with the coefficients from Reunion Islands study.

So, it can be concluded by Figs. 1–3 that the best algorithm for Antananarivo is the Meteo France Algorithm. Up to now, the study will be focused only on the Meteo

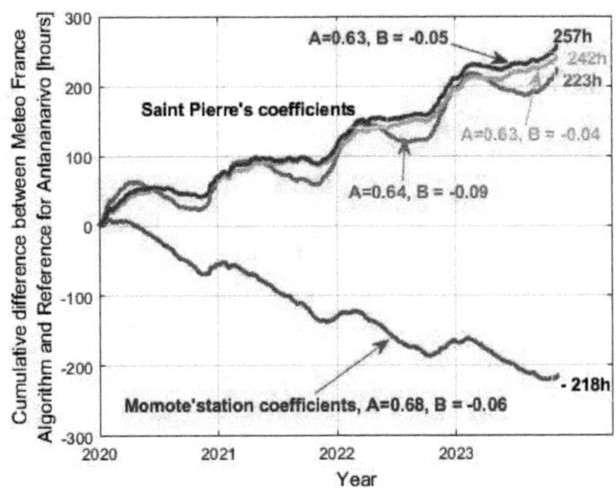

Fig. 4. The plot of the cumulative daily difference between Meteo France Algorithm and the reference.

France Algorithm, and it is this method that will be improved to well fit the data from Antananarivo.

Figure 4 illustrates the plot of the cumulative of daily differences for only the Meteo France Algorithm. As it can be seen, Momote stations coefficients give a continual underestimation of the cumulative daily sunshine duration, whereas the coefficients from Saint Pierre locality yield a continual overestimation. So, for these two cases, the total difference is always increasing by the year. It is then quite logical to think that it is possible to find coefficients, such that results will fall between the two groups, in order to have a curve that is as close as possible to zero.

4. Improving the Meteo France Algorithm

Improving the Meteo France Algorithm means finding the suitable couple of A and B coefficients in order to get the minimum uncertainties. As developed in [4–5] it is almost an empirical method. By intuition and after some trials, an interval with a step of 0.01 is defined for each coefficient. Then, the algorithm is running for all possible combinations of the coefficients, and the suitable coefficient is the one that gives the minimum difference with the reference. For Antananarivo, the interval was [0.58; 0.55] for A coefficient, and [−0.15; 0.07] for B coefficient. By a step of 0.01, it gives 18 possible values for A, and 23 values for B, which in total gives 414 possible combinations. Figure 5 gives the total cumulative differences in hours for all these 414 combinations. The best combination is the one where the total cumulative differences are nearest to zero, according to Fig. 5, it should be between the 175th and 200th combinations. Figure 6 gives a zoom for the 175th to 185th combination.

There are two combinations that give the nearest to zero total cumulative differences. The first is for $A = 0.65$, $B = 0.01$ with 0.83 h for total cumulative difference, and the second is for $A = 0.65$, $B = 0.04$ that gives 0.32 h of total

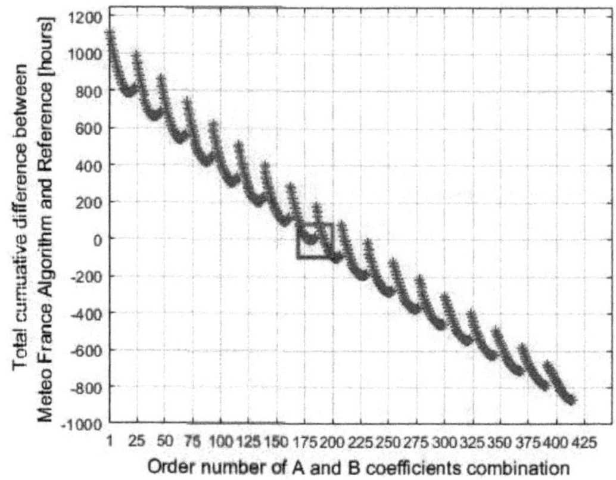

Fig. 5. Total cumulative difference for the 414 possible combinations of A and B coefficients.

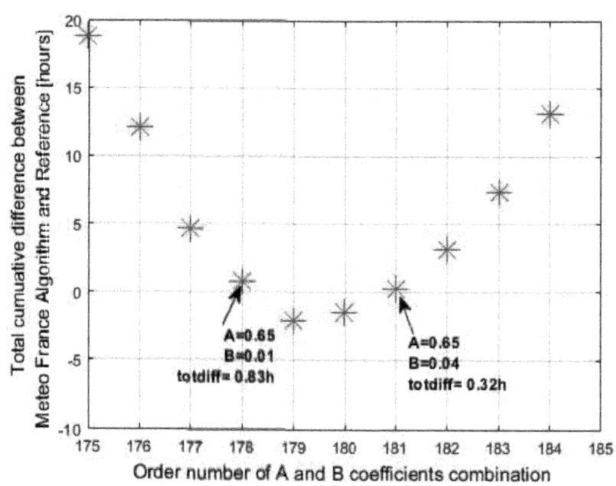

Fig. 6. Zoom of Fig. 5 for 175th and 185th combinations of A and B coefficients.

cumulative difference. Comparing these results to the 218 h of total difference by the coefficients from Momotes station, the present coefficients, are, by far, the most suitable for the present site. Figure 7 gives the plot of the cumulative differences for Antananarivo, with present and previous studies coefficients. Obviously, the new coefficients for the present study are the best for the town of Antananarivo, because the curve is oscillating near and around zero over the four years of measurement, while it is continuously increasing or decreasing for the coefficients from Momotes station or from the previous study for Saint Pierre locality. Table 4 gives the value of the mean daily differences (*mdd*), the standard deviation of the daily differences (*sdd*), and the NRMSE%.

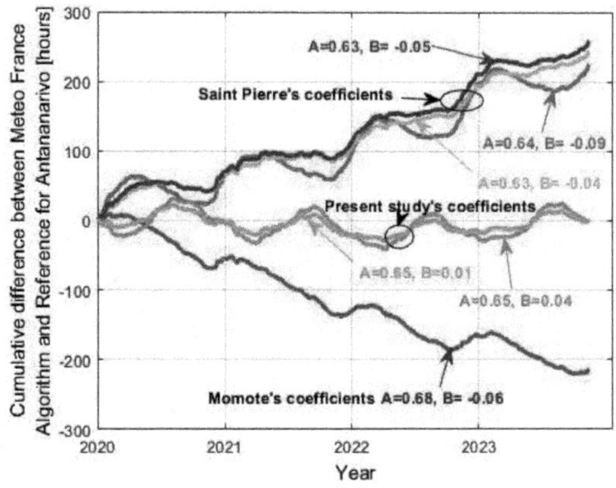

Fig. 7. Plot of the cumulative difference for Meteo France Algorithm with present and previous A and B coefficients.

Table 4. Coefficients for the linear algorithm.

Set of coefficients for Meteo France Algorithm	Mean daily differences (*mdd*) in hour	Standard deviation of the differences (*sdd*) hour	Total cumulative differences in hour	NRMSE (%)
Momote $A = 0.68$, $B = -0.06$	−0.15	0.52	−218	8.42%
Saint Pierre set 1 $A = 0.64$, $B = -0.09$	0.16	0.53	223	11.54%
Saint Pierre set 2 $A = 0.63$, $B = -0.04$	0.17	0.61	243	9.84%
Saint Pierre set 3 $A = 0.63$, $B = -0.05$	0.18	0.63	258	10.2%
Presents study set 1 $A = 0.65$, $B = 0.01$	0.0006	0.55	0.83	8.54%
Presents study set 2 $A = 0.65$, $B = 0.04$	0.0002	0.62	0.32	9.65%

For the mean daily differences and the total cumulative differences, the coefficients from the present study give the best result, and the difference is very big compared to the previous coefficients. There are 0.0002 h of mean differences compared to the 0.15 h from the previous coefficients. A comparison of 0.32 h of total cumulative differences for the present studies coefficients with the 218 h from the previous coefficients is studied. But for the standard deviation of the differences and the NRMSE%, there is no significant difference between the previous coefficients and the present study's coefficients. The present study gives two set of coefficients; to compare between them, Fig. 8 shows a zoom of Fig. 7, where the plot is focused on the two new set of coefficients.

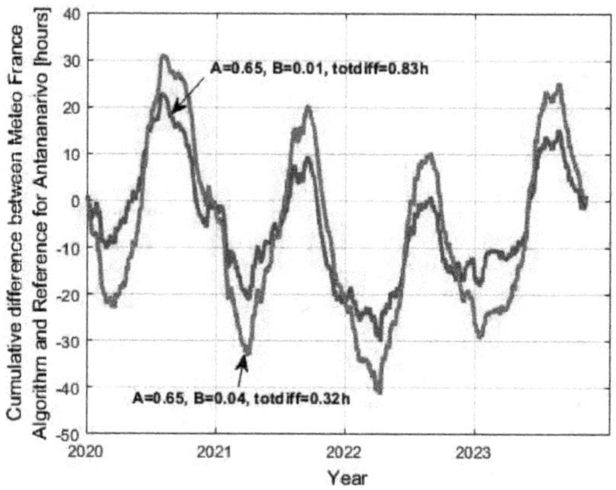

Fig. 8. Plot of the cumulative difference for Meteo France Algorithm for the present studies coefficients.

The best set is the one whose plot is closest to the zero axis. By Fig. 8, the best is the first set ($A = 0.65, B = 0.01$) because the maximum of the negative or positive cumulative peak difference is always lower than for the second set ($A = 0.65$, $B = 0.04$). This is also emphasized by the NRMSE% parameter, for the first set the NRMSE% is 8.54% and 9.65% for the second set.

5. Discussion

5.1. *About the best pyranometric algorithm for Indian Ocean Island*

The result of the present study confirms that the best method for Indian Ocean Island is the Meteo France Algorithm. Of course, it is still needed to find the appropriate values for A and B coefficients, but when this adjustment is done, this algorithm gives the best result among the existing. So, from now on, the next study, for other islands, will be focused only on the Meteo France Algorithm.

5.2. *About the improvement of the Meteo France Algorithm*

This improvement as shown in [4-5] is an empirical method. An interval is defined by a step of 0.01 for each coefficient, then the algorithm is run for all possible combinations of the coefficients. There are three categories of test parameters, the first is the long-term behavior by the total cumulative difference, with the mean and standard deviation of the daily differences. The second is the short-term behavior by the NRMSE% test, and the third is the plot of the cumulative difference. If a hierarchy must be done between these parameters, the most important is the plot of cumulative difference, followed by the NRMSE%, and the mean and standard deviation of the daily differences. For the suitable A and B coefficients, the plot of the cumulative difference is a curve that oscillates near and around zero, whereas for the other set of coefficients, it is continual increasing or decreasing curves. If the curve of two or more sets of coefficients is oscillating near and around zero, the best is the one in which the negative or positive peak difference is the lowest. But the improvement does not change too much the NRMSE%, the value is similar for all sets of coefficients.

5.3. *Future work*

This paper is one step among many others inside a global program to study the estimation of the sunshine duration from the global radiation in a horizontal plane for the islands in the Indian Ocean. So, the next work is to continue the study, for the all other islands, using only the Meteo France Algorithm, and to find a conclusion for the suitable coefficients for the region. On the other hand, the Slob & Monna Algorithm should be revisited to well fit tropical site in Indian Ocean. The main parameters to study are the estimation of Linke Turbidity, direct beam irradiance, and diffuse component.

6. Conclusion

The best method to estimate the sunshine duration from solar global radiation for tropical site, case of the town of Antananarivo is the Meteo France Algorithm. The main coefficients of this algorithm have been evaluated, improved and adjusted for the data base. Comparison based on statistical method between the present work results and the references data shows a good agreement. The other algorithms are inappropriate, in their actual form, for tropical site unless and until they will be revisited and corrected to fit data from tropical site.

ORCID

T. Fanjirindratovo https://orcid.org/0009-0003-2131-4447

References

1. World Meteorological Organization, **8**, I.8-1 (2008).
2. Y. B. L. Hinssen, Comparison of different methods for the determination of sunshine duration, *Royal Netherlands Meteorological Institute Report (KNMI)*, **WR 2006-06** (2006) pp. 72.
3. J. C. Oliviri, *World Meteorol. Org. IOM Rep.* **877**, 357 (1998).
4. E. Vuerich, J. P. Morel, S. Mevel and J. Oliviri, Updating and development of methods for worldwide accurate measurements of sunshine duration, in *Proc. TECO*, Brussels, Belgium, 16–18 October (2012) pp 1–22.
5. T. Fanjirindratovo, D. Calogine, O. Chau and O. Ramiarinjanahary, *Solar* **4**, 455 (2024).
6. E. Vuerich, J. P. Morel, S. Mevel, F. Foti, T. Iorios, P. Cucchiarelli and G. Daddario, Measurements methods for sunshine duration (SD): Accuracy evaluation of pyranometric methods and SD measuring instruments, in *Proc. TECO*, Helsinki, Finland, 30 August–1 September (2010).

7. Y. B. L. Hinssen and W. H. Knap, *J. Atmos. Ocean. Technol.* **24**, 835 (2007).
8. W. H. Slob and W. A. Monna, *Tech. Rapporten* **136**, 31 (1991).
9. J. P. Morel, Sunshine duration measurements using the Carpentras method, in *Proc. Baseline Surface Radiation Network Meeting*, Postdam, Germany, 1–3 August (2012).
10. ENERGY-Lab University of Reunion Island, https://galilee.univ-reunion/fr/thredds/catalog/dataTextfiles/Madagascar/antananarivo/catalog.html.
11. J. J. Michalsky, *Sol. Energy* **40**, 227 (1988).
12. K. N. Liou, *An Introduction to Atmospheric Radiation*, 2nd edn. (Academic Press, Cambridge, MA, USA, 2002).
13. C. N. Long and E. G. Dutton, BSRN global network recommended QC tests, V2.0. BSRN Tech. Rep., https://epic.awi.de/30083/1/BSRN/recommended_QC_tests_V2.pdf (2022).

Grid-connected PV inverter stability via H_∞ and μ-analysis*

R. Georges,[†,‖] A. G. Rabearivelo,[‡]
H. I. Rajaosolomanantena[§] and O. Ramiarinjanahary[¶]

[†]*Doctoral School of Biodiversity and Tropical Environments, University of Toliara, Madagascar*

[‡]*Automatic Systems Engineering, Polytechnic Higher School of Antananarivo, Madagascar*

[§]*Doctoral School of Science and Technology for Engineering and Innovation (ED STII), University of Antananarivo, Madagascar*

[¶]*Physics and Environment Laboratory, University of Toliara, P.O. Box 185, Toliara 601, Madagascar*
[‖]*georgesribiere1@gmail.com*

Received 17 March 2025
Revised 14 August 2025
Accepted 21 August 2025
Published 30 October 2025

In automation theory, the generic dynamic model of the three-phase inverter connected to the electrical grid depends on the operation of the insulated-gate bipolar transistor (IGBT), with a diode mounted antiparallel. In order to have synchronized output voltage and frequency on the generator side, (even if the system is disturbed by the direct effect of weather and loads), uncertain parameters and external disturbances must be accurately measured when the solar photovoltaic (PV) source is interconnected to the JIRAMA's power grid. To ensure the system's robust stability, we recommend defining the system's dynamic performance criteria. So, the aim of this study is to present all the key steps in the realization of the equivalent electronic circuits of the three-phase inverter according to the modes of operation of the IGBT transistors with a diode in parallel (equivalent to the switch); then, to have all the voltage and current equations of the system calculated in order to obtain the state and observation equations of the overall system before modeling the detailed schematic of the three-phase inverter; it is very important to prove the dynamic performances by displaying the values of the poles of the system.

Keywords: IGBT transistor; modeling; dynamic performance; poles values.

*Talk given at the 14th International Conference in High-Energy Physics - HEPMAD24, 21–26th October 2024, Antananarivo-Madagascar.

‖Corresponding author.

1. Introduction

The inverter connected to JIRAMA's power grid, with its high penetration of solar photovoltaic (PV) sources, is very sensitive to load variation and to the effect of changes in the sunshine rate on the production side, which leads to the risks of voltage fluctuation and variations in the inverter's output currents. In particular, it is the full-wave operation that we will be adapting to our study strategy, as it improves voltage quality, boosts system efficiency and ensures component longevity. The behavior of a three-phase inverter must be mathematically modeled for the voltage and current. The modeling and control of this three-phase system is indeed nonlinear. The control law ensures linearity and the study of dynamic performance criteria and system stability. It is emphasized that the transfer function model of the nominal system is infelicitous due to the uncertain parameters. As the PV system is interconnected to the grid, it is essential to use the full-wave modulation technique, which is qualified to analyze the operation of the three-phase inverter switch. The use of the controllability technique and observation allows us to the study system stability by calculating the values of the system poles. In this paper, we will present the equivalent circuits and equations of state, including a detailed block diagram of the system; then we will demonstrate the dynamic performance criteria so that we can confirm that this system meets the required conditions; and, before concluding, it is necessary to specify the values of the poles to prove that the system meets the stability condition.

2. Representation of System Equations of State

Reference 1 published, announced the use of the DC/AC conversion technique using Pulse Width Modulation (PWM). This technology can accommodate various inverters, such as Voltage-Source Inverters (VSI) and Current-Source Inverters (CSI). Figure 1 is subdivided into four parts, of which : the production source is considered the PV generator; r_c, C and I denote the resistor in series with the DC bus capacitor, which maintains the stable operating point and performs the function of energy accumulator and grid-connected inverter; r_{Lk} and L_k ($k = a, b, c$) denote the filter circuit

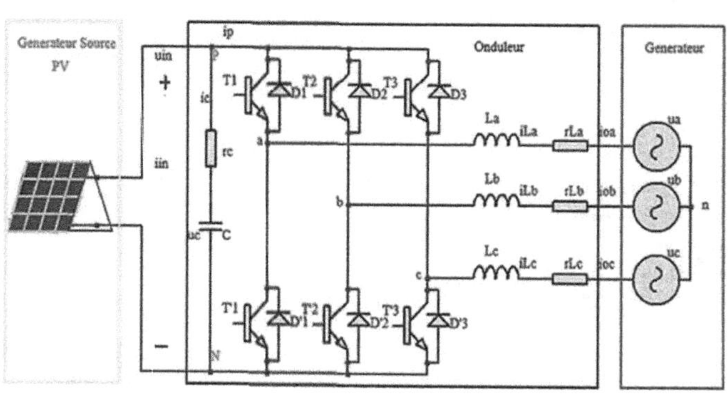

Fig. 1. Three-phase grid-connected solar PV système.

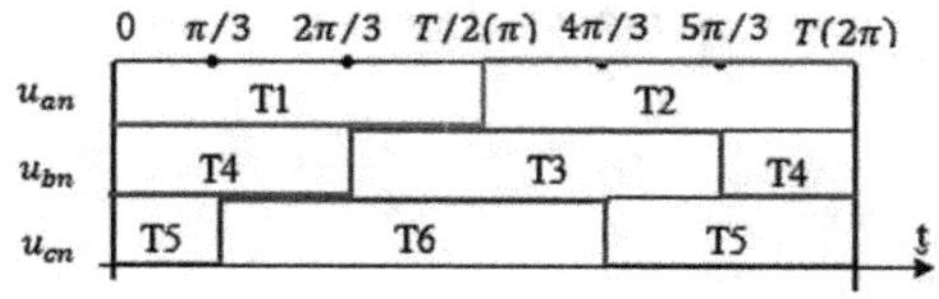

Fig. 2. Strategy diagram for the full-wave control method.

which acts to reduce ripple components due to switching actions in full-wave control schemes and, finally, u_a, u_b, u_c are, respectively, the three-phase network voltages.

Articles 1 and 2 demonstrate the importance of interconnecting the VSIs between the PV system and the power grid. Indeed, the effects of system instability that occur during the power transmission process lead to voltage fluctuation and current distortion.

First, we will demonstrate the mathematical model of the system in the case where the action of the power supply switches of the electronic power converter works in the ideal state, and follows the complementarity states, which are defined in articles 3 and 4 as follows:

$$T_i = \begin{cases} 1, & \bar{T}_i = 0 \\ 0 & \bar{T}_i = 1 \end{cases} \quad \text{for } i = 1, 2, 3 \tag{1}$$

By applying the full-wave modulation technique, we can simplify the study of the mathematical model of the three-phase system (Fig. 2). First, we are going to make the complete equivalent circuits of the system at each step of the modulation cycle in a period T, of an interval from [0 to 2π]. Next, we will apply Kirchoff's principle to circuit meshes and apply the law of nodes to circuit currents and voltages. Thus, we will adopt this strategy in order to facilitate the interpretation of the relationship between the input and output equations of the three-phase inverter using the state model of the system with input vectors and output vectors.

For this reason, the operation of the three-phase inverter is summarized in the following table: Table 1 shows the operation of switches starting with the binary values $V_{b0} = 000$ and ending with $V_{b7} = 111$, as shown in Fig. 2.

Figure 1 interprets switch complementarity in binary values 000 to 111, which are nonzero switch states. Furthermore, from these two binary values, we are able to present the complete equivalent circuits of the system in the case where the switches work in closed and open states over the time interval $[0 - 2\pi]$. The system is assumed to be in a steady-state condition. To develop a mathematical model, it is important to represent the three-phase inverter using equivalent inverter switch diagrams, as shown in Fig. 3–8.

Table 1. Three-phase inverter operation and the full-wave method.

T1T′2T3 V_{b3} et V_{b6}	T1T′T′3 V_{b1} et V_{b4}	T1T2T′3 V_{b2} et V_{b5}	T′1T2T′3 V_{b6} et V_{b3}	T′1T2T3 V_{b4} et V_{b1}	T′1T′2T3 V_{b5} et V_{b2}
101 010 0 à $\pi/3$	100 011 $\pi/3$à $2\pi/3$	110 001 $2\pi/3$à π	010 101 πà $4\pi/3$	011 100 $4\pi/3$à $5\pi/3$	001 110 $5\pi/3$à 2π

2.1. *Star circuit with complementary switches*

- **Mode 1:** $\omega t = [0 - \frac{\pi}{3}]$

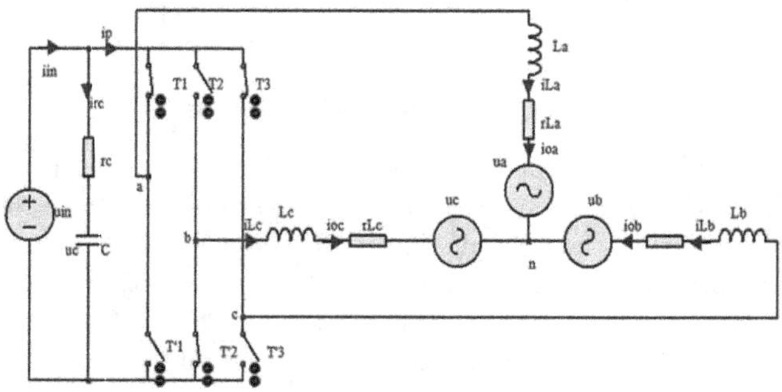

Fig. 3. Equivalent inverter switch diagram T1T'2T3.

- **Mode 2:** $\omega t = [\frac{2\pi}{3} - \frac{4\pi}{3}]$

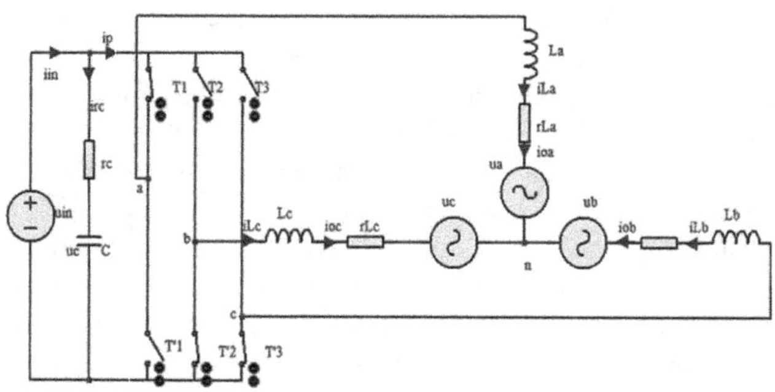

Fig. 4. Equivalent inverter switch diagram T1T'2T'3.

- **Mode 3:** $\omega t = [\frac{4\pi}{3} - 2\pi]$

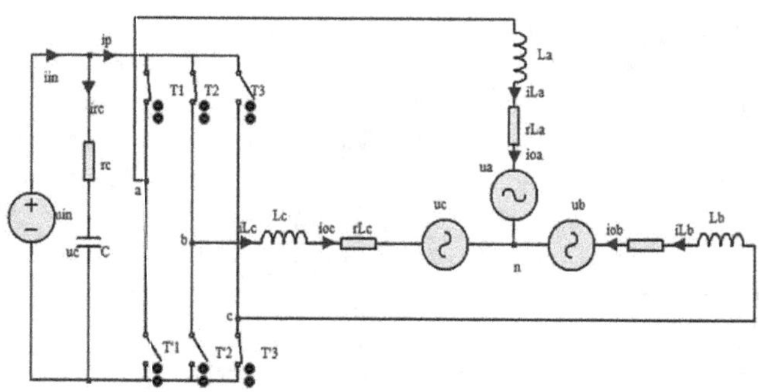

Fig. 5. Equivalent inverter switch diagram T1T2T'3.

- **Mode 4:** $\omega t = [\pi à\ 4\pi/3]$

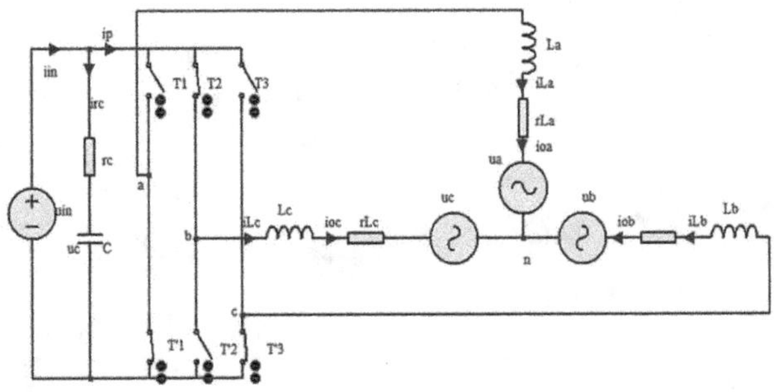

Fig. 6. Equivalent inverter switch diagram T′1T2T′3.

- **Mode 5:** $\omega t = [\frac{4\pi}{3} - \frac{5\pi}{3}]$

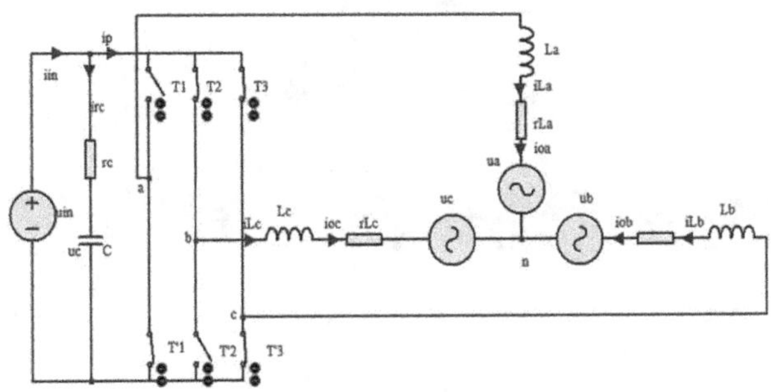

Fig. 7. Equivalent inverter switch diagram T′1T2T3.

- **Mode 6:** $\omega t = [\frac{5\pi}{3} - 2\pi]$

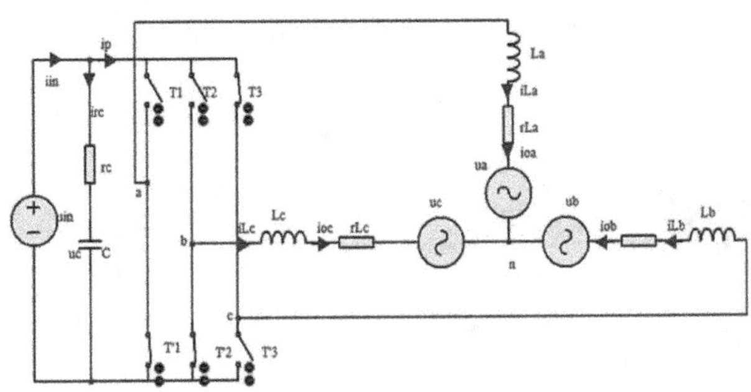

Fig. 8. Equivalent inverter switch diagram T′1T′2T3.

In the following section, we will go into a little more detail about the equations for the currents and voltages of the equivalent system.

2.2. *Equivalent circuit for each switching period*

For each case of the inverter's equivalent circuits, the system input voltage is equivalent to:

$$u_{\text{in}} = u_{rc} + u_c \tag{2}$$

- 1st cas : $\omega t = [0 - \frac{2\pi}{3}]$

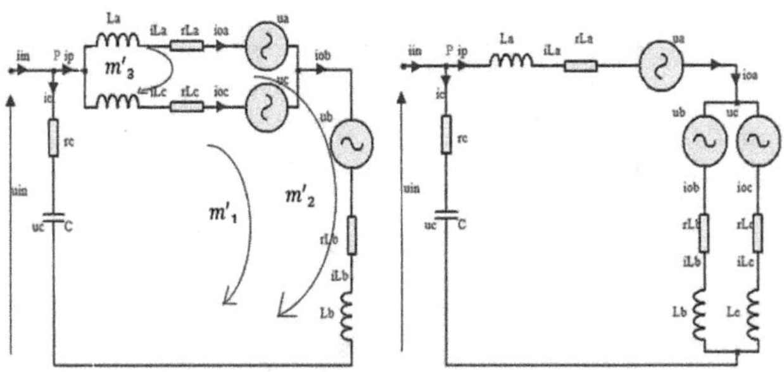

Fig. 9. Equivalent circuit for each switch: $V_{b6}V_{b3}V_{b1}V_{b4}$.

- 2nd cas : $\boldsymbol{\omega t} = [\frac{2\pi}{3} - \frac{4\pi}{3}]$

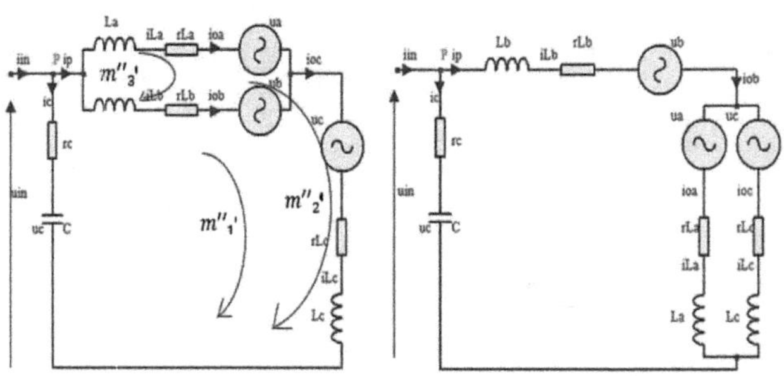

Fig. 10. Equivalent circuit for each switch: $V_{b2}V_{b5}V_{b3}V_{b6}$.

- 3rd cas : $\boldsymbol{\omega t} = [\frac{4\pi}{3} - 2\pi]$

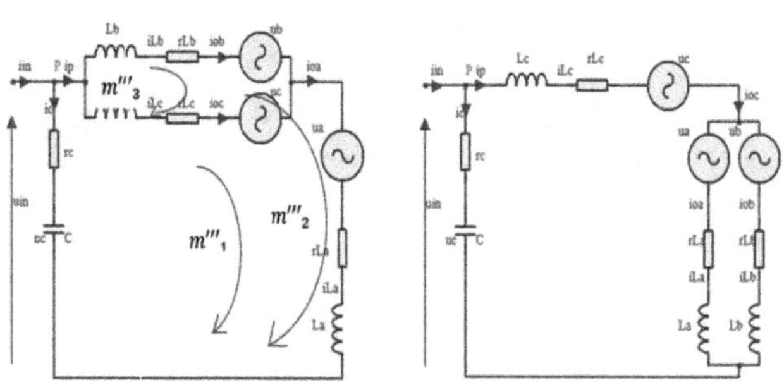

Fig. 11. Equivalent circuit for each switch: $V_{b4}V_{b1}V_{b5}V_{b2}$.

2.3. *Equivalent circuit voltage equations for each case*

By applying Kirchoff's law of star circuit coupling, we can derive the equations for each stage in $V_{b6}V_{b3}$, $V_{b1}V_{b4}$, $V_{b2}V_{b5}$, $V_{b3}V_{b6}$, $V_{b4}V_{b1}$ and $V_{b5}V_{b2}$.

2.4. *Equivalent circuit current equations for each case*

Note that for each case of equivalent circuits, the current law gives us representations (of the system currents) equal to:

$$i_{\text{in}} = i_p + i_c \tag{3}$$

The study of the stages in $V_{b6}V_{b3}$, $V_{b3}V_{b6}$ and $V_{b5}V_{b2}$ systems focuses on the three-phase inverter, but the behavior and the dynamic properties differ when the switches are switched. So, since the condition has already been fixed, we can write the current inputs capacitor and the mean value equations of the inverter three-phase modelas as follows :

$$i_c = i_{\text{in}} - i_{\text{ob}}, \tag{4}$$

$$i_c = i_{\text{in}} - i_{\text{ob}}, \tag{5}$$

$$i_c = i_{\text{in}} - i_{\text{oc}} \tag{6}$$

and

$$i_{\text{oa}} = i_{La}, \tag{7}$$

$$i_{\text{ob}} = i_{Lb}, \tag{8}$$

$$i_{oc} = i_{Lc}. \tag{9}$$

Plant stability depends on a robust, high-performance regulation system.

3. Mathematical Model of a Three-Phase Inverter

The behavior of a three-phase inverter must be modeled mathematically. The modeling of this three-phase system is, in fact, required to ensure dynamic performance criteria and system stability.

3.1. *Solving voltage fluctuation problems*

The inverter connected to JIRAMA's Toliara-I power grid, with its high penetration of solar PV sources (hereinafter referred to as PV), is highly sensitive to load variation and to the effect of changing sunshine levels on the production side. This hazard creates the risk of voltage fluctuations and variations in the output currents of the three-phase inverter[1] and frequency variation. In fact, the system's precision is unstable in the face of a threat from outside; a threat that can be seen as a modification of the internal system. In this progress report, we will present the equations of state and the detailed functional modeling, followed by the dynamic performance criteria, so that we can confirm that this system can meet the required conditions.

3.2. *Mathematical modeling of the system in the abc normal reference frame*

We are going to select a multivariable process describing the operation of an inverter by applying the optimization algorithm. In this section, we focus on modeling a control system to control the voltage and current of the three-phase inverter to be studied. By applying Kirchoff's principle, the system dynamics of a three-phase inverter is summarized in the modeling equation of state and the observation equation, according to Figs. 9–11:

$$\begin{bmatrix} \frac{du_C}{dt} \\ \frac{di_{La}}{dt} \\ \frac{di_{Lb}}{dt} \\ \frac{dL_C}{dt} \end{bmatrix} = \begin{bmatrix} \dot{u}_C \\ i_{La} \\ i_{Lb} \\ i_{Lc} \end{bmatrix} = \begin{bmatrix} 0 & -\frac{1}{3C} & -\frac{1}{3C} & -\frac{1}{3C} \\ \frac{1}{2L} & -\frac{(-rC+2La)}{2La} & 0 & 0 \\ \frac{1}{2Lb} & 0 & -\frac{(-rC+2Lb)}{2Lb} & 0 \\ \frac{1}{2L} & 0 & 0 & -\frac{(-rC+2Lc)}{2Lc} \end{bmatrix} \begin{bmatrix} u_C \\ i_{La} \\ i_{Lb} \\ i_{Lc} \end{bmatrix} + \begin{bmatrix} \frac{1}{C} & 0 & 0 & 0 \\ \frac{rC}{2L} & \frac{1}{La} & 0 & 0 \\ \frac{rC}{2Lb} & 0 & \frac{1}{Lb} & 1 \\ \frac{rC}{2Lb} & 0 & 0 & \frac{1}{Lc} \end{bmatrix} \begin{bmatrix} i_{in} \\ u_{an} \\ u_{bn} \\ u_{cn} \end{bmatrix}, \tag{10}$$

$$\begin{bmatrix} u_{in} \\ i_{La} \\ i_{Lb} \\ i_{Lc} \end{bmatrix} = \begin{bmatrix} 1 & \frac{-r_C}{3} & \frac{-r_C}{3} & \frac{-r_C}{3} \\ 0 & 1 & 0 & 0 \\ 0 & 0 & 1 & 0 \\ 0 & 0 & 0 & 1 \end{bmatrix} \begin{bmatrix} u_C \\ i_{La} \\ i_{Lb} \\ i_{Lc} \end{bmatrix} + \begin{bmatrix} r_C & 0 & 0 & 0 \\ 0 & 0 & 0 & 0 \\ 0 & 0 & 0 & 0 \\ 0 & 0 & 0 & 0 \end{bmatrix} \begin{bmatrix} i_{in} \\ u_{an} \\ u_{bn} \\ u_{cn} \end{bmatrix}. \tag{11}$$

In the stationary three-phase coordinate frame (abc) of the system, the internal mathematical representation of a grid-connected inverter can explicitly describe its behavior through the dynamics of its internal variables. We then obtain equations of state for voltages and currents according to the Kirchhoff's principle for the grid-connected inverter, of the following form:

$$\begin{cases} \dot{x}_{Lk} = A_{uciok}x(t) + B_{inuk}u(t) \\ y = C_{uciok}x(t) + D_{inuk}u(t) \end{cases}. \tag{12}$$

All representations of each parameter A_{uciok}, B_{inuk}, C_{uciok} and D_{inuk} are presented in the detailed block diagram, which is in line with the reality of the evolution of technological progress in research.

4. Dynamic System Performance and Stability

First of all, the transfer function from $U(s)$ to $Y(s)$ in Eqs. (10) and (11) can be easily verified using the Laplace transform:

$$G_s(s) = D + C(sI - A)^{-1}B. \tag{13}$$

The matrix quadruplet $(ABCD)$is a minimal realization of the $G(s)_s$ eigentransfer function if it satisfies the following assumptions:

$$(\text{A1})(A, B) \text{ is commandable}: \ \text{Com} = \text{rang}(B, AB, \ldots, A^{n-1}B) \in \mathbb{R}^n \tag{14}$$

$$(\text{A2})(C, A) \text{ is observable}: \ \text{Obs} = \text{rang}([C, CA, \ldots, CA^{n-1}]^T) \in \mathbb{R}^n \tag{15}$$

So, in the case of our system, the criteria are well met, since the value of $Rank\ Com = 4$; this implies that the pair (A, B) is controllable. And likewise for the value of $Rank\ Obs = 4$; this explains why the pair (C, A) is observable.

Hence, it is from the eigenvalues of A that we can define the asymptotically stable and unstable systems. Indeed, according to the calculation, we obtain the values of the poles of the model are $p_{11} = 10e - 04 * (-0.0054 + 0.8544)$, $p_{12} = 10e - 04 * (-0.0054 + 0.8544)$, $p_{21} = 10e - 04 * (-0.0109 + 0.0000)$ and $p_{22} = 10e - 04 * (-0.0109 + 0.0000)$.

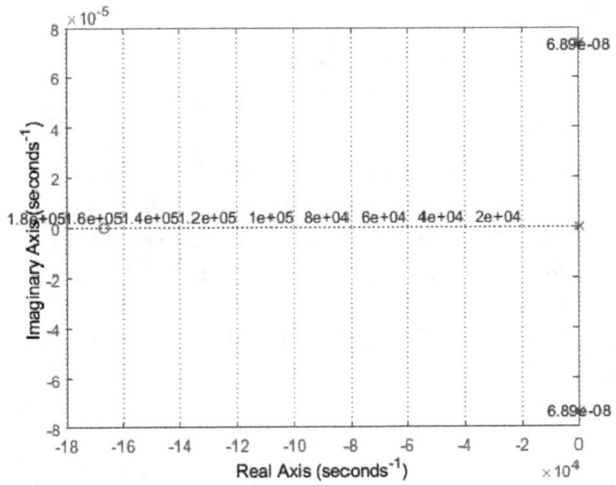

Fig. 12. Placement of poles.

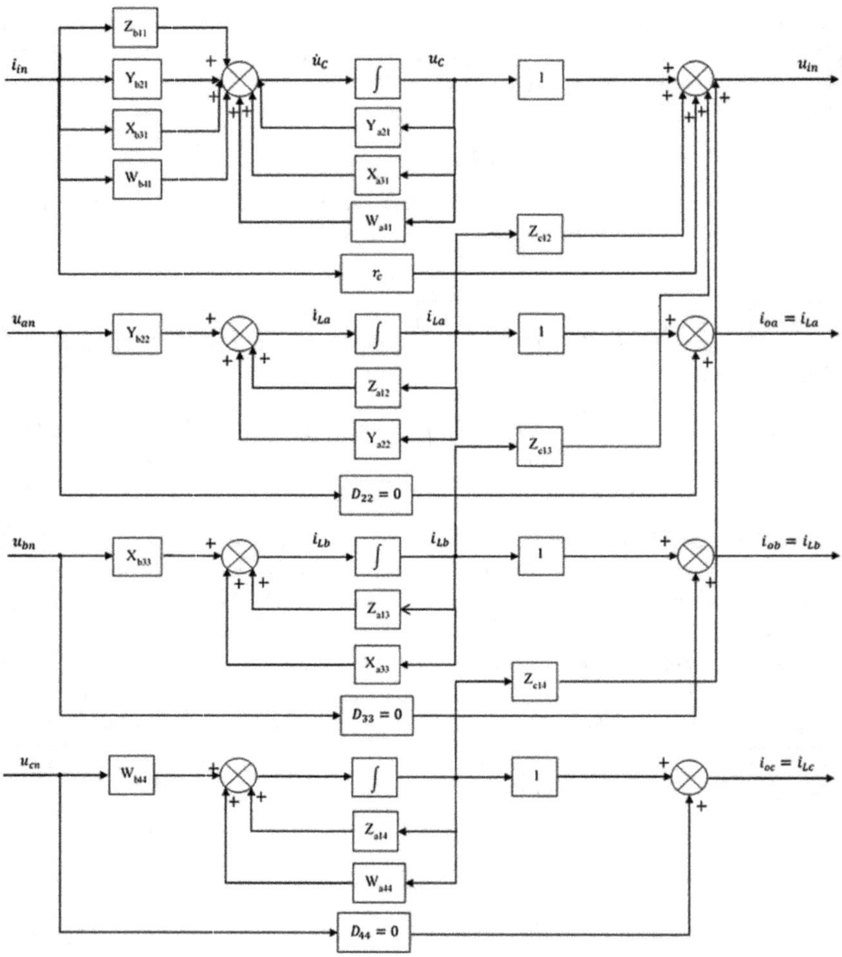

Fig. 13. Three-phase inverter system block diagram.

So, depending on the nature of the poles, we can see that model stability is guaranteed, as all four poles are from the negative reals. And here is an illustration of the position of the poles:

Compared with the value of the poles in Fig. 12, we can see that it satisfies the stability criteria.

5. Conclusion and Perspective

This paper explores an important contribution to the realization of equivalent circuits of the three-phase inverter of the solar PV system interconnected to the

JIRAMA power grid based on the operation of switches during the period [0 to T] or [0 to 2π] using 180°control. The importance of these circuits has enabled us to carry out all the current and voltage equations to form the equations of state, and to observe the whole three-phase system. By using these equations, we were able to validate that the system met the performance criteria for controllability and observability, which are essential before testing the system's stability. Next, the location of the pole values in the negative real part of the complex plane allowed us to conclude that the system is stable. And finally, the three-phase inverter system modeling is well done. The system stability is not necessarily efficient and robust. So, at this stage, we are curious to continue analyzing the system's stability and robustness performance using the H_∞ controller technique and μ-analysis under the influence of structured and unstructured disturbances and uncertainties. Hence, in order to linearize the system, the H_∞ controller is designed by choosing weighting functions W_1 and W_2 to improve the performance of the voltage and the frequency of the nominal unstable system.

References

1. F. L. Luo and H. Ye, *Advanced DC/AC Inverters: Application in Renewable Energy* (CRC Press, 2012).
2. T. Yang, Z. Cai and Q. Xun, *IEEE* **8** (2020).
3. J. Pukko and T. Suntio, *IET Renew. Power Gener.* **6**(6), 381 (2012).
4. R. Georges, A. G. Rabearivelo, O. Ramiarinjanahary and M. Justin, Contribution to the modeling of the stability of the JIRAMA network interconnected with a photovoltaic power source in the Atsimo-Andrefana, *13th High-Physics International Conference*, Organized by HEPMAD from 23–28 October 2023, https://www.slac.stanford.edu/econf/C221102/.

Part 6

Chemical Physics

Chairs: T. Fanjirinjatovo,
V.G. Randrianandrasana

X-ray fluorescence characterisation of clays from Toliara*

Veloson Grégoire RANDRIANANDRASANA†, Dahazizy SETA, Rabetsitonta Jacquelin TSIMIHAFY

Faculté des Sciences, Université de Toliara, BP 185, Toliara 601, Madagascar

Received Day Month Year
Revised Day Month Year
Accepted Day Month Year
Published Day Month Year

When it comes to making traditional bricks in Toliara, clay material specialists and scientists with expertise in the field are not consulted, and it would appear that their advice is invaluable if we are to benefit more from better bricks, while the sellers are only satisfied that there are buyers for their product, without any knowledge of the scientific aspects. This study highlights the need for in-depth physico-chemical analysis of Toliara clays, located in the south-west region of Madagascar, in order to reveal the intrinsic chemical composition of the material. In this study, the X-ray fluorescence(XRF) spectrometer was used to analyze ten clay samples collected from a specified distance in the locality of Ankoronga in the rural community of Betsinjake.Prior to analysis, the samples were pulverized to a size of 80 µm and then subjected to a steam treatment. The analysis showed that the moisture content ranged from 2.13 to 8.85%, indicating the presence of some montmorillonite. The device detected seventeen elements in oxide form, with concentrations expressed as a percentage. These elements are Mg, Al(7.484 to 28.450%), Si(52.386 to 73.560%), K(1.563 to 3.064%). In addition, Ca, Ti, V, Cr, Mn, Fe, As, Sr, Zr, Nb, Mo, Sn and Sb were detected in varying concentrations. In addition, the SiO_2/Al_2O_3 ratio(1.918 to 9.465) reveals the presence of interstratified minerals and T-O, T-O-T type minerals, attributed to the kaolinite and montmorillonite groups. In addition, the Al_2O_3/Fe_2O_3 mass ratio(4.295 to 11.083) suggests that these samples are suitable for use in building materials and ceramics.The silicon content, in excess of 50% by mass, indicates that the samples analyzed are clays. It should be noted that these results should be confirmed by the use of other, more powerful instruments, which would reveal the possible presence of other minerals in these samples, and that an increase in the number of samples analyzed could lead to improved results.

Keywords: Clay; XRF; characterization; Toliara.

PACS numbers: 03.65.−w, 04.62.+v

*Talk given at the 14th International Conference in High-Energy Physics - HEPMAD24 (21-26th October 2024, Antananarivo-Madagascar).
†Speaker: enamboare@yahoo.fr

1. Introduction

The Akoronga brickworks, located at the entrance to the town of Toliara, is renowned for the manufacture of traditional clay bricks. However, scientific information on the physico-chemical characterization of this clay was limited. It was with this in mind that this research was carried out, with the aim of providing solid scientific data accessible to all. In Madagascar, the manufacture of artisanal clay bricks requires no binders or chemicals; each brick is made entirely from clay.[1] Indeed, clay comes from soils or sedimentary rocks, which mainly contain finely dispersed minerals such as phyllosilicates(or sheet silicates), silicas(colloidal silica, quartz) and iron and aluminum oxy-hydroxides.[2] Clay is distinguished by its remarkable physico-chemical properties, including malleability, adhesiveness, thermal conductivity, slip and flow properties, and high mineral and trace element content. These characteristics make clay a material of choice in multiple fields of application, including health, wellness, beauty, personal care, home care, construction and industry.[3]

2. Methods

This study focuses on the physico-chemical characterization of these clays, aiming to identify the chemico-chemical elements of the clay. In this study, core sampling was used to collect ten samples from a single clay deposit, selected at a certain distance. These samples, taken from depths ranging from half to two and a half metres, are distinguished by two distinct colors: brown and greyish. Prospecting required the use of a smartphone equipped with a GPS application to obtain the geographical coordinates of the study area and to visually document the deposit. Prior to analysis, a mechanical preparation stage was undertaken, consisting of drying the samples, both solid and wet, in an oven at 105℃ for 24 hours, in order to determine the moisture content of the contents. In order to obtain a powder size of less than 200 mesh(80 µm), the samples were pulverized. Each sample, weighing 4 grams, was placed in the dish for a 120-second analysis. The measurement was carried out using a portable X-ray Fluorescence(XRF) spectrometer from Madagascar's Office des Mines Nationales et des Industries Stratégiques(OMNIS).[4]

3. Geographical location

The study area, known as Ankoronga, is located around 15 kilometers south of the city of Toliara, in the southwest region of Madagascar. It lies 3 kilometers from the airport, along the RN7 national road, east of the Mozambique Channel, in the rural commune of Betsinjake, district of Toliara II. The village is located at the following geographical coordinates: Latitude: $23°23'55.5''$ south, Longitude: $43°44'313.3''$ east, corresponding to an altitude of 5 m above sea level. According to the Koppen-Geiger classification, the region's climate is hot and dry desert all year round. According to the Koppen-Geiger classification, the climate is tropical semi-arid.[5] The figure 1 shows the geographical context.

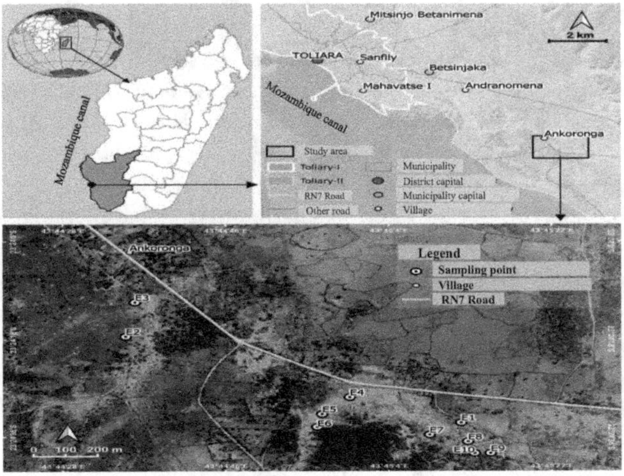

Fig. 1. Geographical map of study area.

4. Geological context

The study area, the subject of this analysis, is located on the southwest coast of Madagascar and belongs to the Morondava sedimentary basin.[6] With the exception of the eastern part of the Mahafaly massif, which belongs to the basement, the Toliara region occupies the southern part of this sedimentary basin. The basin is a succession of north-south trending depressions, slopes(cuestas) and plateaus(reverses).[5] The figure 2 shows the geological map of the site.

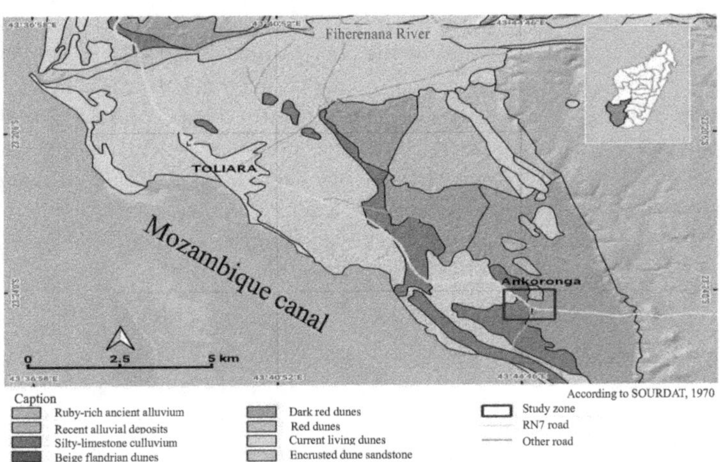

Fig. 2. Geological map of the study area

5. Results and discussion

The ten clay specimens studied, collected by X-ray fluorescence spectrometry, revealed the presence of seventeen distinct minerals. Analysis of each sample revealed the presence of the chemical elements that make up clay, including silicon(52.386 to 73.560%) and aluminum(7.484 to 28.450%), as well as secondary elements. The presence of calcium(8.725 to 18.304%), zirconium(2.140 to 5.311%), potassium(1.563 to 3.064%), iron(1.194 to 2.890%) and magnesium(0.001%) was found to be significant. It should be noted that these elements present a considerable mass content, with concentrations in excess of 1%, with the notable exception of magnesium, whose concentration remains stable in all samples analyzed.Furthermore, the SiO_2 concentration, which amounts to 50% in commercial kaolin originating from Great Britain,[7] is also found in samples E5(54.580%) and E8 (52.386%), classified as kaolinite. Furthermore, the aluminum oxide (Al_2O_3) content of 19.030% in Algerian montmorillonite[8] agrees with the amount of Al_2O_3 in montmorillonite-type samples E4(18.678%), E7(18.503%), E9(18.158%) and E10(19.523%).[9] The instrument also detected the presence of other minerals in minute concentrations, including titanium, vanadium, chromium, manganese, strontium, niobium, molybdenum, tin and antimony. The material balance test shows results between 99.130 and 100.848%, with a margin of error attributable to potential sample contamination and humidity, due to unexpected analysis time. In addition, the presence of organic matter, such as plant debris, as well as chemical elements undetectable by FRX, often with atomic number $Z < 11$, and elements in gaseous state, could also influence results. Kaolinite samples, with an aluminum content of less than 25%, are particularly suited to the building materials industry(brick, tile, etc.)[10] In addition, analysis of samples E3, E1 and E2 reveals the presence of interstratified minerals, as indicated by SiO_2/Al_2O_3 values of 6.360, 8.708 and 9.465, respectively.[11] Furthermore, the values 1.918, 2.320, 2.420 of samples E5, E8, E6 and 3.100, 3.250, 3.260, 3.290 of samples E9, E1 reveal a variation in the composition of the minerals present. 0, E4, E7 of this SiO_2/Al_2O_3 ratio are closer to kaolinite and montmorillonite clays, respectively.[10,12] Furthermore, the Al_2O_3/Fe_2O_3 ratio, which exceeds 5.500, attests to the compatibility of these samples, characterized by a high alumina content, with the manufacture of refractory ceramics. On the other hand, samples with a ratio of less than 5.500, due to their high iron content, are recommended for the production of refractory ceramics. In this results table the samples are numbered from E1 to E10.(See Table 1)

Table 1. Analysis results of clay samples

Minerals	Samples									
	E1	E2	E3	E4	E5	E6	E7	E8	E9	E10
MgO	0.001	0.001	0.001	0.001	0.001	0.001	0.001	0.001	0.001	0.001
Al_2O_3	7.484	7.895	11.235	18.678	28.450	24.863	18.503	22.569	18.158	19.523
SiO_2	65.181	73.560	71.420	60.920	54.580	60.053	60.928	52.386	56.111	60.184
K_2O	2.451	2.964	2.579	2.631	1.563	3.064	2.322	2.637	2.568	1.654
CaO	17.007	10.161	11.207	12.752	10.312	8.725	12.142	18.304	18.119	10.180
TiO_2	0.616	0.494	0.433	0.616	0.418	0.482	0.560	0.546	0.415	0.641
V_2O_3	0.017	0.012	0.013	0.017	0.015	0.017	0.018	0.015	0.013	0.641
Cr_2O_3	0.018	0.016	0.017	0.018	0.017	0.014	0.021	0.019	0.017	0.015
MnO	0.044	0.034	0.036	0.044	0.077	0.072	0.049	0.074	0.060	0.059
Fe_2O_3	1.742	1.264	1.194	1.742	2.566	2.896	2.119	2.556	2.064	2.353
As_2O_3	-	-	-	-	0.001	0.001	0.001	0.001	0.001	0.001
SrO	0.007	-	0.004	0.007	-	0.015	-	0.007	-	-
ZrO_2	4.810	3.804	2.417	3.015	2.140	-	4.053	-	3.225	5.311
Nb_2O_5	0.004	0.004	0.002	0.004	0.003	0.003	0.004	0.003	0.003	0.003
MoO_3	0.001	0.001	0.001	0.001	0.001	0.001	0.001	0.001	0.001	0.001
Sn_2O_5	0.012	0.002	0.011	0.012	0.002	0.010	0.001	0.011	0.007	0.009
Sb_2O_3	0.013	0.013	0.013	0.013	0.010	0.006	-	-	0.010	0.006
$\sum$ (%)	99.408	100.225	100.583	100.848	100.155	100.222	100.721	99.130	100.771	100.580
SiO_2/Al_2O_3	8.708	9.465	6.360	3.260	1.918	2.420	3.290	2.320	3.100	3.250
Al_2O_3/Fe_2O_3	4.295	6.246	5.790	8.740	11.083	8.600	7.790	8.830	8.800	8.300
Humidity level (%)	2.13	7.96	8.85	7.50	7.60	8.23	7.98	7.43	4.07	4.94

6. Conclusion

As part of this study, the ten clay samples were analyzed using a portable x-ray fluorescence spectrometer(FRX). This instrument detected seventeen chemical elements in oxide form, including Mg, Al, Si, K, Ca, Ti, V, Cr, Mn, Fe, As, Sr, Zr, Nb, Mo, Sn and Sb. In addition, the contents of major elements such as silicon(52.386 to 73.560%) and aluminum(7.484 to 28.450%) were determined. The SiO_2/Al_2O_3 ratio(1.918 to 9.465) corresponds to interstratified minerals and T-O, T-O-T type minerals, attributed to the kaolinite and montmorillonite groups. Furthermore, the Al_2O_3/Fe_2O_3 mass ratio (4.295 to 11.083) of these samples is recommended for the manufacture of building materials(bricks, tiles, etc.) and ceramics. However, it has been found that this type of brick performs less well than the local parpaing, but its availability at an affordable cost contributes to its persistence in construction practices.Consequently, it is essential to continue efforts to improve the quality of local bricks. To achieve this, it is imperative to ensure a balanced silica and alumina content, the use of clean water, free from salt and contaminants, and optimization of the manufacturing process(drying, firing, mechanical pressing, etc.).

ORCID

Veloson Grégoire Randrianandrasana `https://orcid.org/0009-0002-4099-1681`
Dahazizy Seta `https://orcid.org/0009-0009-6689-0097`
Rabetsitonta Jacquelin Tsimihafy `https://orcid.org/0009-0001-6472-0512`

References

1. D. R. Riantsoa, Étude de confort thermique d'un bâtiment en briques artisanales d'argile cuite naturellement ventilé: Détermination de l'orientation optimale, Ph.D. thesis, Université d'Antananarivo (2020).
2. Interaction intercouche et criblage électronique dans le graphène multicouche étudiés¡? format?¿ par spectroscopie de photoémission à résolution angulaire, Lettres d'examen physique 98 206802.
3. F. Hernot, L'argile, son utilisation à l'officine, Diplôme d'Etat de Docteur en Pharmacie, université Angers (2016).
4. D. Ndahirwa, H. Zmamou, H. Lenormand, N. Leblanc, The role of supplementary cementitious materials in hydration, durability and shrinkage of cement-based materials, their environmental and economic benefits: A review, Cleaner Materials 5 (2022) 100123.
5. B. T. Z. Fredo, Les palmiers dattiers phoenix dactylifera à toliara: Étude de la filière, utilisation et diversité variétale.
6. R. Battistini, G. Richard-Vindard, Biogeography and ecology in Madagascar, Vol. 21, Springer Science & Business Media, 2013.
7. D. Wetshondo Osomba, Caractérisation et valorisation des matériaux argileux de la province de kinshasa (rd congo) (2012).
8. I. Belamri, S. Hioul, F. E. Bouremmad, Synthèse de molécule pharmaceutique par catalyse hétérogène en présence de la montmorillonite traitée, Ph.D. thesis, Université de Jijel (2019).

9. A. Abakar, Caractéristiques mécaniques et thermiques de l'argile stabilisée par la gomme arabique et renforcée par la paille de riz, Ph.D. thesis, Université de Lorraine (2018).
10. R. S. Deramne, Elaboration de céramiques silicatées à base des matières premières argileuses et d'un déchet végétal de centrafrique: Caractéristiques physicochimiques et frittage, Ph.D. thesis, Université de Limoges; Université de Lomé (Togo) (2021).
11. M. El Ouahabi, Valorisation industrielle et artisanale des argiles du maroc (2013).
12. A. Qlihaa, S. Dhimni, F. Melrhaka, N. Hajjaji, A. Srhiri, Caractérisation physico-chimique d'une argile marocaine [physico-chemical characterization of a morrocan clay], J. Mater. Environ. Sci 7 (5) (2016) 1741–1750.

X-Ray Fluorescence characterization of igneous salt in the commune of Tameantsoa, South-west madagascar*

Jean-Claude Omer ANDRIANARIMANANA[2], Andriamiarimbola Irène RAKOTONIAINA[3], Miora Koloina RANAIVOSOA[4], Veloson Grégoire RANDRIANANDRASANA[1], Dahazizy SETA[1]† Rabetsitonta Jacquelin TSIMIHAFY[1]

[1] *Faculté des Sciences, Université de Toliara, BP 185, Toliara 601, Madagascar*

[2] *Ecole Normale Supérieure, Université d'Antananarivo, BP 566, Antananarivo 101, Madagascar*

[3] *Faculté de médecine, Université de Toliara, BP 185, Toliara 601, Madagascar*

[4] *Faculté de médecine, Université d'Antananarivo, BP 566, Antananarivo 101, Antananarivo, Madagascar*

Received Day Month Year
Revised Day Month Year
Accepted Day Month Year
Published Day Month Year

In Madagascar, many natural resources are not exploited according the standard such as the fossil salt of the commune of Tameantsoa in the Southwest region where the salt is composed of several minerals in particular the sodium ion (Na^+) and the chloride ion (Cl^-) which are essential for human, animals and environment health. The twelve (12) salt samples were pulverized to have powder of less than 200 mesh (80 μm). The analysis was carried out in the laboratory of the Office des Mines Nationales et des Industries Stratégiques (OMNIS), by X-Ray Fluorescence (XRF). After stoving, the compounds detected were sodium, chloride (68.64 to 85.45%); elements such as magnesium, aluminum, phosphorus, sulfur, potassium, calcium, titanium, vanadium, chromium, manganese, iron, nickel, copper, zinc, arsenic, zirconium, molybdenum, selenium, rubidium, strontium, barium and tungsten. The density and volumic mass calculated were from 2.109 to 2.354 and from 2.106 to 2.351 kg/m^3 respectively as well as the pH found ranging from 6.10 to 6.75. The amount of NaCl was relatively lower compared to its content in sea salt (91%), so it is important for food. The use of more efficient device would contribute to the confirmation of our results and also could highlight the existence of others minerals composition. Increasing the number of samples analyzed would provide

*Talk given at the 14th International Conference in High-Energy Physics - HEPMAD24 (21-26th October 2024, Antananarivo-Madagascar)

†Speaker: seta.dahazizy@gmail.com

a more solid result.

Keywords: Igneous salt; XRF; characterization; diagram-binary; Tameantsoa.

PACS numbers: 03.65.−w, 04.62.+v

1. Introduction

Concentrated in the northern part (Antsiranana), the Menabe region (Morondava) and various towns in the south (Toliara I and II).[1] Salt is also mined in the communes of Tameantsoa, Tongobory and Bezaha, Betioke district, south-west region. This type of igneous salt, known in Malagasy as "SIRATANY", is reputed to be good for health, according to some people, but is less widely sold and consumed than sea salt in the Toliara market. In advanced countries, industrial salt requirements were several times higher than food consumption.[2] In the United States, for example, almost 97% of total production (around 40 million tonnes) is used for non-food purposes.[3] Salt plays an essential role in the human body, maintaining the hydromineral balance. It was involved in regulating blood pressure and volume. Any deficiency is harmful to health. Salt deficiency was, in fact, the cause of disorders ranging from mild malaise to prostration, which can be fatal to the organism of the deficiency is not corrected within a very short time.[4] Salt is composed mainly of the chemical elements sodium (Na) and chlorine (Cl). The sodium ion is essential to life, helping to control the volume of the extracellular medium, the transmission of nerve impulses and the intestinal absorption of certain nutrients.[5–7] Salt's chloride ion aids digestion by maintaining stomach acidity.[8] There are also trace elements that we will determine in our analyzed samples.

2. Materials and Methods

The figure 1 illustrates the salt extraction techniques studied in this work.

2.1. *Characterization*

Characterisation methods used in this study include X-ray fluorescence spectrometry (XRF), pycnometer, 105°C oven, thermometer, pH meter and conductivity meter.

2.2. *Geographical and geological situation*

As illustrated in Figure 2, a Geographical location of the study. Tameantsoa is a rural commune located 138 km from the town of Toliara, in the southwest region of Madagascar, on national road n°10. It is about 69 km from Andranovory (RN7).With the exception of the eastern part of the Mahafaly, which belongs to the basement, the Toliara region, southwest coast of Madagascar, occupies the southern part of the Morondava sedimentary basin.[9]

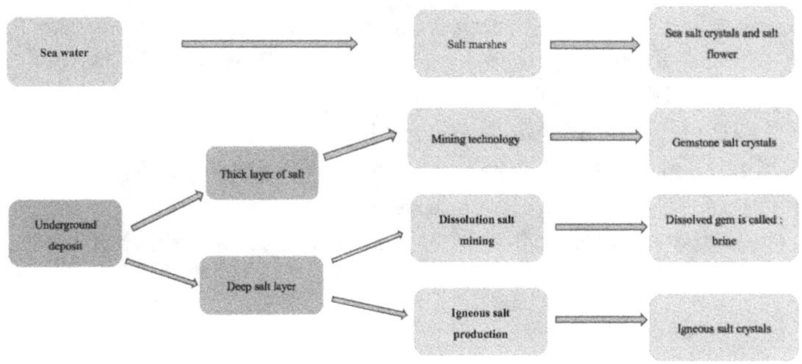

Fig. 1. Salt exploitation process

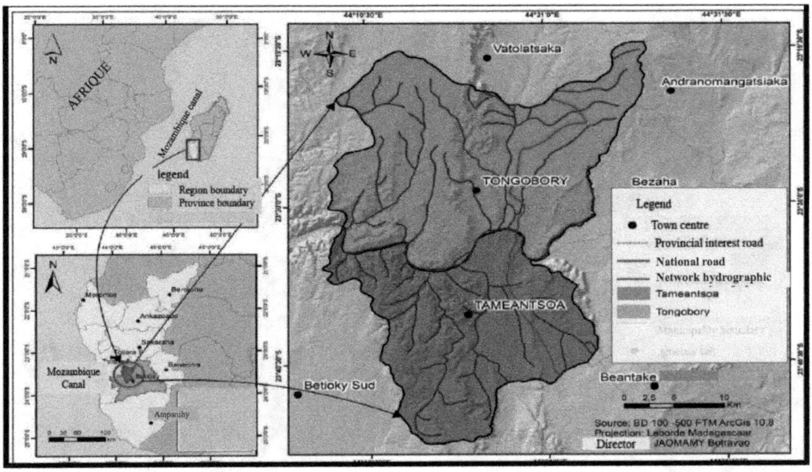

Fig. 2. Geographical location of the study area

3. Results and Discussion

3.1. *Diagrams series*

Changes in the elements Mg+Mn+Fe+Cu+Zn compared with NaCl. The NaCl trend curve (C_2) decreases as the concentration of the chemical elements (Mg+Mn+Fe+Cu+Zn) shown in figure **3a**, increases, so this curve is decreasing for the quantity of these elements, but increasing for the NaCl value (%).
Variation in Ti+V+Cr+Ni elements compared with NaCl. Moreover, at the point

(75.1; 0.01%), the trend line of (C_3) reaches an extremum along the tangent (T), in other words, this curve for NaCl is increasing, while it is decreasing for the chemical elements (Ti+V+Cr+Ni)(Show in figure 3b).
Variations in Al+P+K+S+Ca compared with NaCl. The NaCl trend curve (C_1) increases from the coordinate (69.80; 7.02%) to the point (85.45; 5.6%), meaning that the quantity of the chemical elements (Al+P+K+S+Ca)(Show in figure 3c) decreases as the NaCl quantity of each sample increases, so this NaCl curve is increasing at these points.

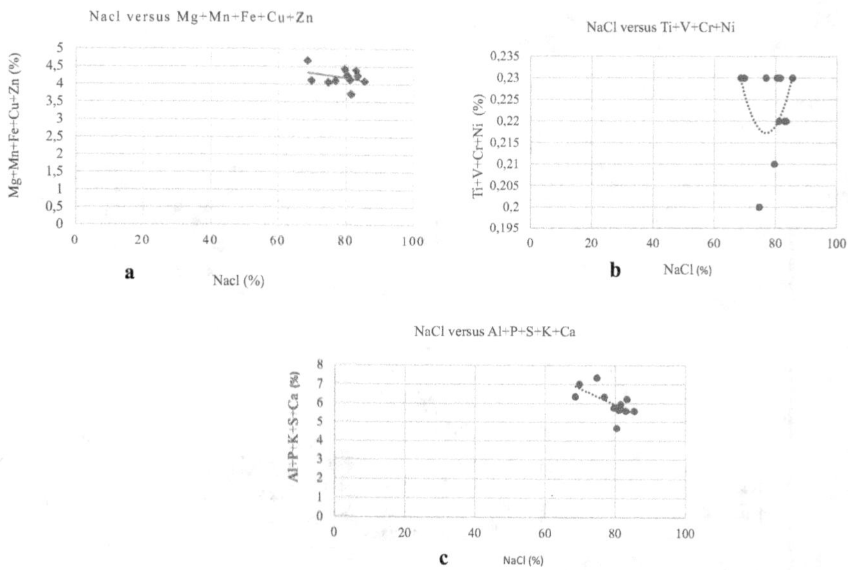

Fig. 3. Binary diagram systems

3.2. *Table results*

Table 1 shows the chemical composition of the samples used in this study. The XRF shows that the sodium chloride (NaCl) content of sample n°10 (85.45%) is comparatively higher than that of sample n°2 (68.64%), which is also comparatively lower than the others. Manganese (Mn), with a concentration of 0.01%, is only found in sample n°3. The concentrations of arsenic (As, 0.01%), a highly toxic element, and undesirable elements such as nickel (Ni, 0.03%) and copper (Cu, 0.01%) are constant across all samples analysed from the sites. The levels of transition metals such as

Titanium (Ti, 0.12%) and Zinc (Zn, 0.01%) remain constant in the twelve samples analysed. The NaCl content exhibited variation, ranging from 68.64% (sample n°2) to 85.45% (sample n°10). It is notable that calcium (Ca) was absent in samples n°4, n°10 and n°12. The quantity of major elements in twelve samples analysed is presented below: Magnesium (Mg) varies from 03.30% to 04.27%, and Aluminium (Al) from 04.10% to 05.54%.

Table 1: Chemical composition

Minerals	**Samples**											
	n°1	n°2	n°3	n°4	n°5	n°6	n°7	n°8	n°9	n°10	n°11	n°12
NaCl	80.43	68.64	79.62	74.76	69.80	83.37	81.49	76.86	82.97	85.45	81.07	83.95
Mg	3.84	4.27	4.01	3.66	3.64	3.84	3.30	3.70	4.00	3.69	3.73	3.68
Al	4.46	5.54	5.23	4.62	5.26	4.73	5.21	4.88	5.34	4.70	4.51	4.10
P	0.11	0.13	0.12	0.08	0.11	0.10	0.09	0.10	0.10	0.10	0.09	0.08
S	0.04	0.67	0.36	2.41	1.64	1.35	0.34	1.14	1.14	0.81	1.04	2.07
K	0.05	–	0.04	0.25	–	0.03	0.31	0.21	0.18	–	–	0.08
Ca	0.01	0.03	0.02	–	0.015	0.02	0.01	0.02	0.01	–	0.01	–
Ti	0.12	0.12	0.12	0.12	0.12	0.12	0.12	0.12	0.12	0.12	0.12	0.12
V	0.02	0.02	–	–	0.02	0.01	0.01	0.01	0.01	0.02	0.01	–
Cr	0.06	0.06	0.06	0.05	0.06	0.06	0.07	0.07	0.06	0.06	0.06	0.06
Mn	–	–	0.01	–	–	–	–	–	–	–	–	–
Fe	0.38	0.38	0.38	0.37	0.45	0.38	0.39	0.38	0.37	0.38	0.38	0.39
Ni	0.03	0.03	0.03	0.03	0.03	0.03	0.03	0.03	0.03	0.03	0.03	0.03
Co	0.01	0.01	0.01	0.01	0.01	0.01	0.01	0.01	0.01	0.01	0.01	0.01
Zn	0.01	0.01	0.01	0.01	0.01	0.01	0.01	0.01	0.01	0.01	0.01	0.01
As	0.01	0.01	0.01	0.01	0.01	0.01	0.01	0.01	0.01	0.01	0.01	0.01
Zr	–	–	–	–	–	0.01	–	0.01	0.01	0.01	–	0.01
Mo	–	–	–	–	–	–	–	0.01	0.01	–	–	0.12
Se	–	0.01	0.01	0.01	0.03	–	0.01	0.01	0.01	–	0.01	–
Rb	0.01	0.01	0.01	0.03	0.01	0.01	0.01	0.02	0.02	0.01	0.03	0.02
Sr	0.02	0.01	0.02	0.01	0.05	0.04	0.03	0.05	0.01	0.03	0.03	0.03
Ba	0.01	0.03	0.02	0.02	0.02	0.02	–	–	0.01	–	0.01	0.01
W	0.05	0.05	0.05	0.04	0.04	0.05	0.04	0.05	0.04	0.04	0.04	0.04
Humidity (%)	9.70	19.97	9.86	13.51	18.58	6.10	8.51	12.28	5.53	4.56	8.80	5.33
$\sum$ (%)	99.37	100.00	100.00	100.00	99.90	100.00	100.00	99.98	99.82	100.00	100.00	100.00

The physico-chemical characteristics of the samples are presented in Table 2 below. Sample n°4 (74.76%) and sample n°5 (69.80%) have different mass concentrations (%) but the same density ($d_4 = d_5 = 2.10$) and Volumic mass ($\rho_4 = \rho_5 = 2.10\,\text{kg/m}^3$). We know that pH_4 (6.12) $\neq$ pH_5 (6.16). Samples n°11 (81.07%) and n°12 (83.95%) also have the same Volumic mass ($\rho_{11} = \rho_{12} = 2.20\,\text{kg/m}^3$) and specific gravity ($d_{11} = d_{12} = 2.20$) but different percentages (%) and hydrogen potential: pH_{11} (6.50) $\neq$ pH_{12} (6.67). Also, the densities ($d_1 = d_3 = 2.26$), Volumic mass ($\rho_1 = \rho_3 = 2.26\,\text{kg/m}^3$) and pH_1 (6.24) $=$ pH_3 (6.24) of samples n°1 and n°3 have the same values but different mass concentrations (80.43% $\neq$ 79.62%).

Table 2. Physico-chemical characteristics

Elements physics (%) / Samples	Density	ρ (Kg/m^3)	pH	Conductivity (mS/cm)
n°1	2.27	2.26	6.24	36.5
n°2	2.20	2.20	6.37	77.1
n°3	2.27	2.26	6.24	39.6
n°4	2.10	2.10	6.12	39.2
n°5	2.10	2.10	6.16	22.2
n°6	2.27	2.26	6.10	33.6
n°7	2.20	2.20	6.75	53.9
n°8	2.27	2.27	6.70	33.7
n°9	2.36	2.36	6.72	38.7
n°10	2.24	2.23	6.75	54.1
n°11	2.20	2.20	6.50	36.4
n°12	2.20	2.20	6.67	24.9

Samples n°1, n°4, n°6 and n°11 have different amounts (grams) of NaCl, but different conductivities and temperatures. Sample n°5 is conductivity of 22.2 mS/cm is lower than that of the others, while sample n°2 (77.1 mS/cm) is the most conductive (see Table 3).

Table 3. Conductivity analysis

Samples	NaCl(g)	NaCl (ml)	Temperature °C	Conductivity(mS/cm)
n°1	2	80	27	36.5
n°2	4	100	27.1	77.1
n°3	3	90	27.3	39.6
n°4	2	70	27.4	39.2
n°5	1	60	27.2	22.2
n°6	2	80	27.4	33.6
n°7	5	80	26.9	53.9
n°8	3	90	27.5	33.7
n°9	4	100	27.4	38.7
n°10	5	80	27.9	54.1
n°11	2	75	27.9	36.4
n°12	1	100	27.9	24.9

4. Conclusions

The samples analysed were obtained from 12 distinct deposits, naturally occurring as minor elevations.Surface samples exhibited higher concentrations of organic matter compared to samples subjected to preliminary analyses, suggesting potential contamination during collection. Additionally, the proportion of NaCl was found to be lower in surface samples due to the washing effect caused by rainwater runoff.The elemental compositions of surface and deep samples were found to be more or less equal, attributed to the low mineralisation levels characteristic of shallow deposits. Furthermore, sodium chloride (NaCl) is an ideal fertiliser for plants, such as coconut palms in Indonesia, where it has been found to have a favourable effect on the general condition of the palms, fruit set and health. In addition, salt is an essential nutrient for animals, with insufficient intake resulting in stunted growth in young animals and fatigue, sluggish milk production and weight loss in adult animals. On modern farms, salt serves as a means for delivering vitamin and mineral supplements, which are essential for the health of livestock.The employment of an additional, more powerful instrument will assist in verifying our findings and may also reveal the presence of other mineral elements.The augmentation of the sample size will engender more robust results.

ORCID

Veloson Grégoire Randrianandrasana `https://orcid.org/0009-0002-4099-1681`
Dahazizy Seta `https://orcid.org/0009-0009-6689-0097`
Rabetsitonta Jacquelin Tsimihafy `https://orcid.org/0009-0001-6472-0512`

References

1. B. Sarrasin, Mining and protection of the environment in madagascar, Mining in Africa: regulation and development (2009) 150–186.

2. S. D. Ohlhorst, M. Slavin, J. M. Bhide, B. Bugusu, Use of iodized salt in processed foods in select countries around the world and the role of food processors, Comprehensive reviews in food science and food safety 11 (2) (2012) 233–284.
3. M. V. Mannar, J. T. Dunn, M. Initiative, I. C. for Control of Iodine Deficiency Disorders, UNICEF, et al., Salt iodization for the elimination of iodine deficiency, ICCIDD, NL, 1995.
4. F. M. Fordyce, Selenium deficiency and toxicity in the environment, in: Essentials of medical geology: Revised edition, Springer, 2012, pp. 375–416.
5. A. Bernal, M. A. Zafra, M. J. Simón, J. Mahía, Sodium homeostasis, a balance necessary for life, Nutrients 15 (2) (2023) 395.
6. H. R. Pohl, J. S. Wheeler, H. E. Murray, Sodium and potassium in health and disease, Interrelations between essential metal ions and human diseases (2013) 29–47.
7. X. Bonneau, D. Boutin, R. Bourgoing, J. Sugarianto, Sodium chloride, an ideal fertilizer for coconut palms in indonesia. (1997).
8. C. Bucking, C. M. Wood, Gastrointestinal processing of na+, cl-, and k+ during digestion: implications for homeostatic balance in freshwater rainbow trout, American Journal of Physiology-Regulatory, Integrative and Comparative Physiology 291 (6) (2006) R1764–R1772.
9. B. T. Z. Fredo, Les palmiers dattiers phoenix dactylifera à toliara: Étude de la filière, utilisation et diversité variétale.

Granulometry analysis of clays from Toliara*

Veloson Grégoire RANDRIANANDRASANA, Dahazizy SETA, Rabetsitonta Jacquelin TSIMIHAFY†

Faculté des Sciences, Université de Toliara, BP 185, Toliara 601, Madagascar

Received Day Month Year
Revised Day Month Year
Accepted Day Month Year
Published Day Month Year

Madagascar has many natural resources that have not yet been legally exploited by large-scale manufacturers, such as clay. Vast deposits of clay are found throughout the south-western region of Madagascar, but their use remains artisanal. The aim of this research is to determine the grain size of Toliara clays in the Ankoronga locality, with a view to transforming this artisanal brick-making site into a modern industry. For this reason, several analyses were carried out to characterize the samples. This began with the determination of clay content and other constituent particles using a granulometric sieve. Particle size analysis indicated a low clay content (4.52 to 11.29%) in this material, with a fineness modulus greater than 2.9. The results showed that these clays, taken in their raw state, do not have all the properties satisfactory for brick production, so further analysis will be required, such as pressure control during firing.

Keywords: Clay; granulometry; sieving; fineness modulus; Toliara.

PACS numbers: 03.65.−w, 04.62.+v

1. Introduction

Since the dawn of mankind, the world has continued to evolve, and the need for shelter has remained constant. Since that time, the aspiration to have one's own reliable shelter to survive in all seasons, and to own something very special(ceramics, tiled roofs, etc.) using raw materials of biological or mineral origin in a variety of techniques, has been common to all. After the appearance of flint and cut stone, clay emerged as the first material used by man in prehistoric times.[1] The unique structure and diverse properties of clays have led to their extensive use in a variety of applications.[1,2] Madagascar is a notable example, with extensive deposits of various clays in all its regions. In Madagascar, clay has been used to produce traditional bricks, which are the most widely used building materials in the country.[3] In

*Talk given at the 14th International Conference in High-Energy Physics - HEPMAD24 (21-26th October 2024, Antananarivo-Madagascar).
†Speaker: trjacquelin@gmail.com

order to contribute to the development and research of this material in the south-west region, with a view to the possible establishment of an industry to transform it into modern building materials such as bricks and tiles, a characterization by granulometry was carried out. The aim of this research was to determine the size of the particles present. The physico-chemical characterization process, using the granulometric sieving method, proved to be influenced by several factors, including the composition of each particle size, quantitative and qualitative analysis, and recommendations for improvement.

2. Materials and Methods

2.1. *Preparation of the sample*

The ten clay samples were analyzed by sieving. They were in a solid, moist state, from ten(10) different sites, of different colors(brown and grayish), and therefore had to undergo the following unit operations in order to obtain representative results and avoid any risk when handling the samples:

- Weighing before and after drying and, in particular, during the analysis process;
- Oven drying for 24 hours(T = 105°C);
- Crushing of samples using a jaw crusher to obtain pieces smaller than 5 millimeter.[4]

2.2. *Equipment*

The measurement was carried out in the laboratory of Madagascar's Office des Mines Nationales et des Industries Stratégiques(OMNIS) in Antananarivo. To facilitate the analysis task, a rot-tape or electric sieve with six series (2500, 500, 250, 200, 100 and 80 µm), from the bottom to the top of the column, was used.[5]

2.3. *Aim of the analysis*

Granulometric analysis is a process which can be used to determine three things: firstly, the grain size; secondly, the proportions of grains of the same size (% by weight); and thirdly, the modulus of fineness(Mf).Sand fineness can be quantified by calculating the Abrams fineness modulus(Mf). This modulus, ranging from 0 to 10, is a quantitative measure of the fineness of the aggregates present in the coarse material. A fineness modulus that approaches 10 indicates a coarse material, while a fineness modulus that approaches 0 indicates a very fine material. It is noteworthy that the higher the fineness modulus, the larger the area that falls above the material's particle size curve.[6]
$Mf = \sum(\text{Cumulative screen rejects}(\%))/100$.
In this work, the quality of raw or fired bricks was recorded after defining the grain size at each sampling site.

2.4. *Particle size analysis protocol*

The sieves are nested one on top of the other, in an order such that the progression of openings increases from the bottom to the top of the column, with six series of sieves (2,500, 500, 250, 200, 100 and 80 µm). A sealed bottom is placed at the bottom of the column to collect the load for later analysis. A cover is placed at the top of the column to prevent any loss of material during sieving. The sample is poured into the upper part of the sieve column, which is then vibrated by a rotating ribbon or electric sieve for 2 to 3 minutes, depending on the load of material to be analyzed. The top sieve and its contents are then removed. The last grains remaining in the sieve must be removed with the utmost care. On a clean tray, the sieve is shaken horizontally, held in one hand and struck against the other (at a rate of around 120 strokes per minute), while brushing the sieve containing fine particles. Never press down on the sieve with the brush, nor forcefully remove a grain trapped in a mesh. Ultrasonic sieve cleaning systems are available. The sieves must not be disassembled, and the column must not be mismatched.[7] Each sieve reject is weighed using a precision balance before being removed from the column.

2.5. *Geographical and geological situation*

Figure 1 below shows the geographical map and the sampling point for each sample in the study area. The ten samples were taken in the Toliara brickworks area located

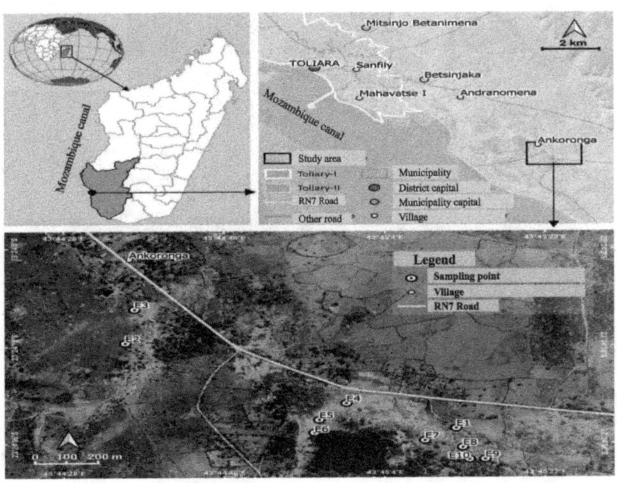

Fig. 1. Geographical location of the study area

in Ankoronga-Andatabo, a small village 3 km from Toliara airport and around 10 Km from the entrance to the town of Toliara on the RN7(national road), east of the Mozambique cannel, in the southwest region of Madagascar. Its coordinates in degrees, minutes and seconds are 23° 23′ 55.5″ South, 43° 44′ 313.3″ East, and its altitude above sea level is 5 meters. With the exception of the eastern part of the Mahafaly, which belongs to the basement, the Toliara region, southwest coast of Madagascar, occupies the southern part of the Morondava sedimentary basin.[8]

3. Results

3.1. *Sieve size curves*

Figure 2 shows the particle size curves for samples n°1(**A**), n°2(**B**), n°3(**C**), n°4(**D**). The analysis curve shows the sieve mesh opening in µm on the x-axis and the cumulative passage rate in % on the y-axis. The passage rate of sample n°1(**A**) on the first sieve (2500 µm) is greater than 80% of the content, while that of sample n°4(**D**) is less than 80%. The clay percentages in these four samples are all below 20%.

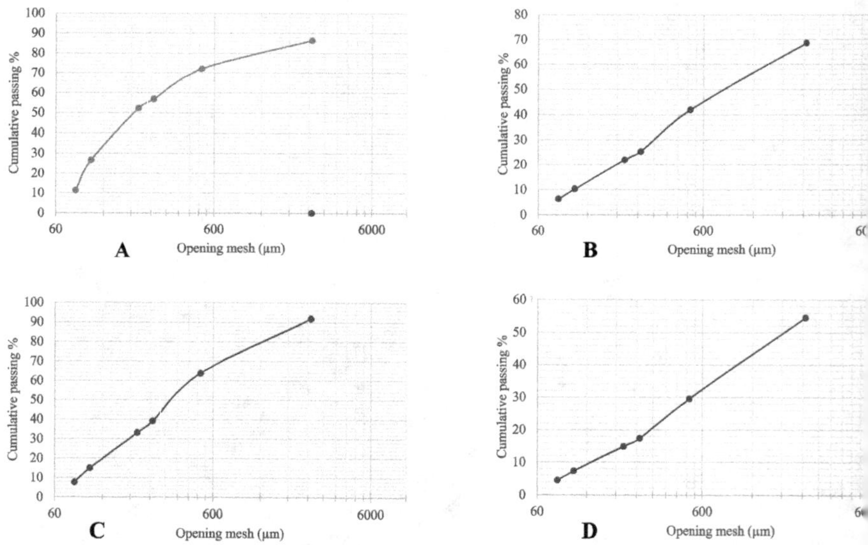

Fig. 2. Sieve size curves for samples n°1, n°2, n°3, n°4

In Figure 3, the percentage of particles passing under the last sieve (80 μm) is less than 10%. All six curves take on an S-shape, although this is weaker for the other samples. As in Figure 2, the alphabetical notation (**E, F, G, H, I, J**) follows the numbers of samples n°5, n°6, n° 07, n°8, n°9, n°10 respectively. Sieve pass varies from 60 to 80%, and the small-grain content is less than 20% in all cases. The curves for samples n° 08 and n°9 are similar in shape, and the final sieve passage, which settles to the Bottom of the sieve, is always less than 10%. The inflection point of these curves is systematically at the 500 to 250 μm level, and the clay content is low in these samples.

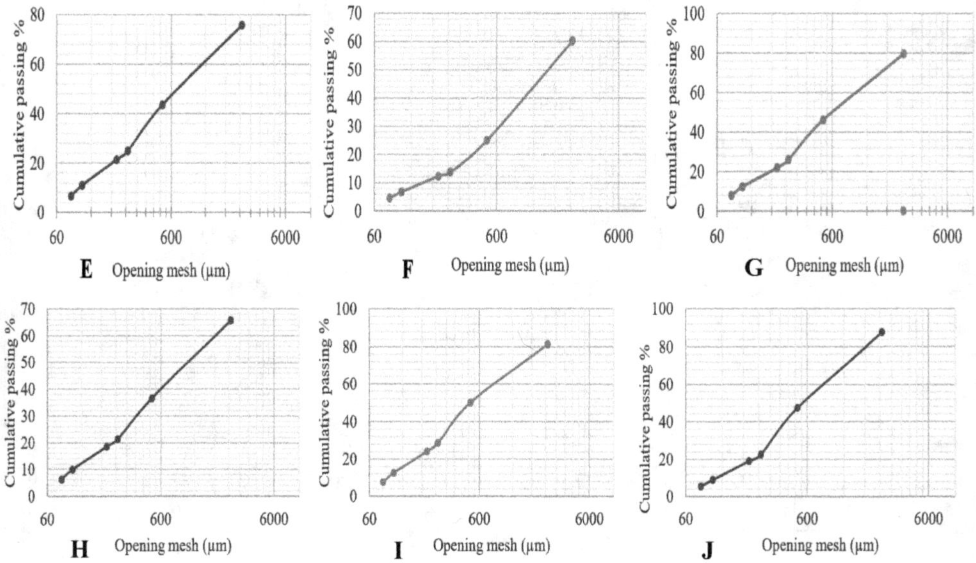

Fig. 3. Sieve size curves for samples n°5, n°6, n°7, n°8, n°9, n° 10

3.2. *Tables of sample analysis results*

The analysis data for each sample are listed in Tables 1 and 2. They include the particle rate in each mesh of the sieve studied, as well as their classification and fineness modulus(Mf).

Table 1. Cumulative passing in the sieve series

Samples	Sieve series (µm) and Cumulative passing (%)					
	2500	500	250	200	100	80
n°1	86.40	71.93	56.72	52.28	26.60	11.29
n°2	68.56	42.06	25.31	21.94	10.25	6.35
n°3	91.88	63.86	39.11	33.22	14.94	7.64
n°4	54.61	29.57	17.40	14.90	7.27	4.52
n°5	75.73	43.44	24.90	21.37	10.99	6.57
n°6	60.25	24.96	13.76	12.32	6.72	4.56
n°7	79.51	46.12	26.00	21.97	12.29	7.92
n°8	65.72	36.60	21.43	18.57	9.92	6.17
n°9	81.34	50.08	28.60	24.24	12.58	7.53
n°10	87.65	47.34	22.59	19.10	9.01	5.41

Table 2. Refusals in the sieve series

Samples	Sieve series (µm) and Reject (%)							Mf
	2500	500	250	200	100	80	Bottom of the sieve	
n°1	13.60	14.48	15.20	4.45	25.67	15.31	11.29	2.95
n°2	31.44	26.51	16.75	3.37	11.69	3.90	6.35	4.26
n°3	8.12	28.02	24.75	5.89	18.28	7.30	7.64	3.49
n°4	45.39	25.04	12.17	2.50	7.64	2.75	4.52	4.72
n°5	24.27	3.28	18.55	3.52	10.38	4.42	6.57	4.17
n°6	39.75	35.29	11.21	1.43	5.60	2.16	4.56	4.77
n°7	20.49	33.39	20.12	4.03	9.68	4.38	7.92	4.06
n°8	34.28	29.12	15.17	2.86	8.86	3.75	6.17	4.42
n°9	18.66	31.26	21.48	4.36	11.67	5.04	7.53	3.96
n°10	12.35	40.31	24.75	3.50	10.09	3.61	5.49	4.09

4. Discussion

According to the data summarized in Tables 1 and 2, all the samples analyzed are not ideal for building materials such as bricks, with the exception of sample n°1, which corresponds to the recommended standard. Indeed, the proportion of clay in these samples is absolutely less than 10%.[9] This means that the sample contains more coarse grains. The modulus of fineness is between 2.95 and 4.77%, so it belongs to the fine class, as these values tend towards zero. Sample n°1 contains a higher proportion of small grains, with a fineness modulus of 2.95%, while sample n°6 includes a higher proportion of large grains, with a fineness modulus of 4.77%. Examination of these tables shows that the Toliara samples are composed of several different particle sizes, such as granules, sand, silt and clay. After rejection on each series of sieves, the classification of the grains can be interpreted as follows:

- 2500 μm : The rejection rate varies between 8.12 and 45.39%, with a higher proportion of granules in sample n°4 than in sample n°3;
- 500 μm : The refusal rate ranges from 14.48 to 40.31%, with the majority of refusals consisting of coarse and medium sand;
- 250 and 200 μm:The rejection rate is between 12.64 and 28.25% and is classified as fine sand;
- 100 μm : The quantity of coarse sand varies from 5.60% to 25.67%. ;
- 80 μm : The rejection rate displayed by this sieve ranges from 2.16 to 15.31%, classifying it as fine sand
- Rejection is classified as silt, mud and clay if the percentage is between 4.52 and 11.29%.

5. Conclusions

The Ankoronga clay deposit near the town of Toliara was chosen to collect samples of different colors. The study focused on granulometric analysis by sieving. The results of this analysis showed that the ten samples analyzed contained less than 11% clay, a percentage below the norm for bricks, and a higher proportion of fine particles. In addition, the study showed that the brick quality obtained by the other samples was poorer than that of sample n°1, whose fine particle content was higher. Further analysis is required to determine structural elements, chemical formula, porosity, density, plasticity and specific surface area.

ORCID

Veloson Grégoire Randrianandrasana `https://orcid.org/0009-0002-4099-1681`
Dahazizy SETA `https://orcid.org/0009-0009-6689-0097`
Rabetsitonta Jacquelin TSIMIHAFY `https://orcid.org/0009-0001-6472-0512`

References

1. L. A. Mango-Itulamya (2015).
2. F. Hernot, *Diplôme d'Etat de Docteur en Pharmacie, université Angers* (2016).
3. N. Liewig, M. Rautureau and C. Gomes, *Etude et Gestion des Sols* **19**, 267 (2012).
4. A. Qlihaa, S. Dhimni, F. Melrhaka, N. Hajjaji and A. Srhiri, *J. Mater. Environ. Sci* **7**, 1741 (2016).
5. J. Juget, *Hydrobiologia* **65**, 145 (1979).
6. J. Fournier, C. Bonnot-Courtois, R. Paris and M. Le Vot, Analyses granulométriques. principes et méthodes (2012).
7. B. GOUBA, *Revue Internationale de la Recherche Scientifique (Revue-IRS)* **1**, 21 (2023).
8. B. T. Z. Fredo.
9. R. G. Elenga, *Annale des Sciences et Techniques* **2** (2019).

Bionanocomposite Films based on Carrageenan and Graphite: elaboration and characterization*

Maherizo T. GEDICE[2], Heriniaina T. JOYEUX[1], Olivier D. MAKA[2], Ernest O.L.J. MAMIARISITRAKA[1], Rabearison M. MILLER[1,2]† Bricharles T. OUDINOT[1], Heriarivelo RISITE[1,2]

[1] *Laboratory of Physics and Environment, Physic Department*
University of Toliara PO Box 185, Toliara, Madagascar

[2] *Test and Analysis of Materials, Institut Halieutique et des Sciences Marines(IH.SM), University of Toliara*
BP 141 Rue Dr. Rabesandratana, Mahavatsy II, Toliara 601, Madagascar

Received Day Month Year
Revised Day Month Year
Accepted Day Month Year
Published Day Month Year

This work involves the development and characterization of new composite materials with biodegradable polymer matrices and graphite reinforcement. First, graphite was exfoliated using the mechanical ultrasound exfoliation method, which is a priori the simplest method for producing graphene sheets in water, followed by the extraction of carrageenans from two red seaweed: Kappaphycus Alvarezii (Cottonii) and Eucheuma Denticulatum (Spinosum). The carrageenan-graphite mixture was then prepared using a solution mixing and casting-evaporation technique, with the aim of producing bionanocomposite films. This was achieved by dissolving carrageenan and dispersing graphite in water, then homogenizing the mixture and spreading it on a suitable surface via a natural drying process. Finally, the new materials obtained were characterized by Thermogravimetric analysis(TGA), X-ray Diffraction(XRD) and Fourier Transform Infrared(FT-IR). Very promising results were obtained from the analysis of the TGA, XRD and FT-IR spectra, highlighting the important role played by particular graphite in obtaining bio(nano)composites with excellent performance and properties.

Keywords: Red seaweed; carrageenan; graphite; graphene; vasting-evaporation; bio(nano)composites.

PACS numbers: 03.65.−w, 04.62.+v

*Talk given at the 14th International Conference in High-Energy Physics - HEPMAD24 (21-26th October 2024, Antananarivo-Madagascar).
†Corresponding author: rabearisonmiller@gmail.com

1. Introduction

In the current context, we are witnessing a period of rapid and profound change affecting all aspects of society. This transition, characterised by technological advances and increased energy consumption, has given rise to a number of challenges that threaten the integrity of our environment. In response to these concerns, researchers and industry leaders are working to identify more efficient materials that can meet both market demands and environmental commitments.[1,2] This strategy consists of substituting more destructive materials with other, less degrading alternatives, without harming the environment.[3] In Toliara, there are many sources of materials that can be used as reinforcements, including graphite, to expand the areas where seaweed can be used.[4] The bonds within the graphite planes are covalent, while the inter-planar bonds are of the Van der Waals type (weak bonds), which makes graphite easy to exfoliate.[5,6] An isolated sheet of graphite is called "graphene", which was first isolated(mechanical exfoliation) in 2004 by Andre Geim and Konstantin Novoselov.[7] Over the past twenty years, with its exceptional properties in a single material, graphene has a strong, hard quality that can be used in bionanocomposites.[8–10] Since 1989, Madagascar has been among the countries cultivating and producing red seaweed such as Cottonii and Spinosum, cultivating mainly the territorial zone and the South-West region, and continues to this day with the support of the University of Toliara and international partners join forces to make village aquaculture a real lever for sustainable development.[11] Global production of marine macro-seaweed has more than tripled since the turn of the millennium, rising from 10.6 million tonnes in 2000 to 32.4 million tonnes, with Madagascar producing 9.665 tonnes in 2019.[12] These red seaweed are among the algae with high levels of polymers after extraction also known as "carrageenans".[13,14] Their main structural differences lie in the position and number of sulfate groups, and the presence or absence of the 3,6 anhydrogalactose(DA) unit.[15] A distinction is made between κ-carrageenan (22% sulfate), ι-carrageenan (32% sulfate) and λ-carrageenans (38% sulfate).[16] κ-carrageenan is soluble in water, but insoluble in organic solvents and oils. The ι-carrageenanhas is partially soluble when cold and totally soluble when hot.[17] The aim of this work is to exploit carrageenans extracted from seaweed by developing new composite materials. It aims to develop high performance bio(nano)composite films using the casting-evaporation method, to study their morphological and structural properties, and to help evaluate the potential benefits of these materials.

2. Materials and Methods

The raw materials used in this work were red seaweed: Kappaphycus alvarezii or Cottonii and Eucheuma denticulatum or Spinosum, donated by Ocean Farmers. The graphite powder used in this work was received from Fotadrevo deposit, Ampanihy District, Atsimo Andrefana Region (South coasts of Madagascar), Toliara.
As illustrated in Fig. 1, a specific strategy was employed to produce the bionanocom-

posite film under investigation.

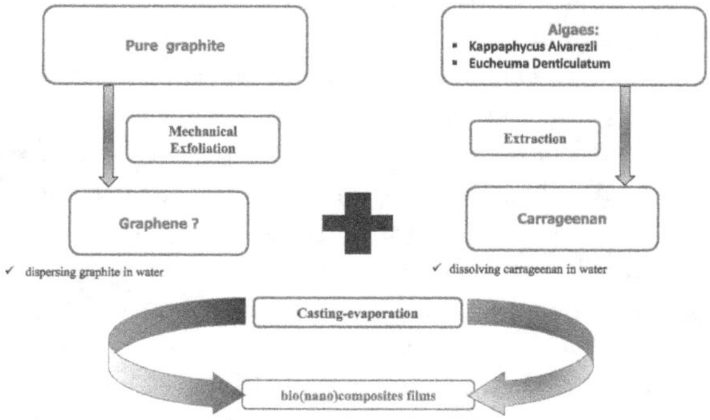

Fig. 1. Strategy for preparing bio(nano)composite films

2.1. *Extraction of carrageenan*

The 5 grams of dry seaweed cut into small pieces were tremerized for 12 hours in a beaker containing 200 ml (milliliter) of distilled water. After 12 hours of rehydration, the distilled water used was now replaced by 40 ml of Sodium Hydroxide solution(N_aOH) and 160 mL of distilled water. At the same time, the Memmert-type thermostatic bath was heated to 90°C to accommodate the prepared samples. When the required temperature(90°C) was reached, the beakers were placed in the bath for 4 hours. After 4 hours of heating, the samples are removed from the Marie bath and sieved through a 125 μm fine mesh filter. After sieving, 500 ml of 90° ethanol was added to the filtrate to precipitate the carrageenans. This step separated the desired carrageenans from the rest of the samples. Alcohol-insoluble carrageenans were manually collected in Petri dishes using a spatula. The precipitates were then recovered using a sieve and stored in petri dishes. Finally, leave the carrageenans to dry in the sun for at least 72 hours.[18]

2.2. *Film production using the casting-evaporation technique*

Firstly, the quantity of graphite powder (0.05 and 0.1 grams, i.e. 1 and 2%, respectively) was dissolved in 200 ml of distilled water. Concurrently, the ultrasonic bath was heated to 80°C in order to receive the prepared graphite solution.Subsequent to the graphite being in solution, exfoliation could begin. This was achieved by subjecting the graphite solution to 60 minutes of sonication in an ultrasonic bath, followed by shearing every 10 minutes using an immersion blender.The application of shear (extrusion) ensures exfoliation by progressively breaking or detaching the graphite sheets. The intensity of the shear then encourages the polymer chains to intercalate in the interfoliar space, exerting local stresses that promote the exfoliation mechanism. The solution mixing of graphite and carrageenan used in this work is made possible by the ability to disperse graphite and dissolve carrageenan in water. Dispersing fine powders in water with agitation and ultrasound can transform graphite into graphene. The mixture of polymer and graphite was prepared at different percentages (by mass) using the solution mixing method followed by the casting-evaporation approach. All blends are prepared with 75% carrageenan and 25% glycerol, with or without filler (by mass). While the graphite is being ultrasonically exfoliated, the carrageenan (3.75 grams, 75%) is dissolved in a beaker containing 100 ml of distilled water under magnetic stirring at a temperature of 40-60 °C with the glycero (1.25 grams, 25%). Following sonication, the carrageenan and glycerol were mixed and poured into the graphite suspension. The resulting mixture was then heated to 80°C and stirred for 45 minutes to homogenise the solution and ensure good dispersion of the graphite in the matrix (carrageenan). Finally, the resulting mixture was spread onto plastic supports (Petri dishes), which were placed in air at room temperature to evaporate the water. During evaporation and drying, the Petri dishes were covered with parchment paper to avoid any effect of evaporation on the mixture and then left at room temperature for 4 to 5 days to ensure complete evaporation of the water.

2.3. *Characterizations*

Three characterisation methods are employed in this study: thermogravimetric analysis (TGA) on an STD Q600 V8.1 Build 99 (Instrument TA), X-ray diffraction(XRD) measurements using a Panalytical Empyrean XRD and Fourier transform infrared analysis(FT-IR) on a Vertex 70V.

3. Results and Discussion

3.1. *Thermogravimetric analysis(TGA)*

Fig. 2 show the mass loss curves for carrageenan from spinosum(Carr-Sp) and cottoni(Carr-Co).The trends in the curves are similar, but the degradation temperatures(T_d) are different. Vaporization of the water molecules is at 62 °C, while total decomposition of Carr-Co is at 242 °C. For Carr-Sp, the water molecule vaporizes at 82 °C, with significant mass loss at 229° C. This is attributed to the loss of physically adsorbed water and the decomposition of Carr-Sp.[19,20]
For the Sp/Gr (carrageenan from spinosum) films(Fig. 3a), the trends in the curves are still similar but they differ in the carrageenan decomposition temperatures, which are 268 °C for $Sp/Gr1\%$ and 278 °C for $Sp/Gr2\%$. This implies that the addition of fillers increased the temperature of carrageenan degradation. This finding suggests that graphenes are present, which was the objective in the composites, as graphites alone would not give such an improvement. Ref. 21 showed that the addition of graphite to the polymer brought an improvement of only 5 °C,

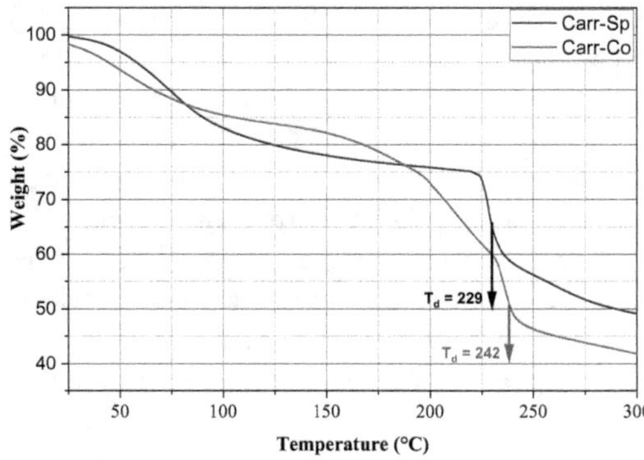

Fig. 2. TGA curves of Carr-Co and Carr-Sp

whereas in most cases the addition of graphenes can greatly improve the decomposition temperature of polymers. Ref. 22 showed that the degradation temperature of nanocomposites with a graphene filler is plus 40 °C compared to pure polymer. Other studies also show that a nanocomposite containing 1% graphene degraded at a temperature of around 40-50 °C compared with pure polymer.[23] We can therefore say that graphene exists in composites. In the case of the composite with 2% filler, the difference in decomposition temperature is around 10 °C. According to TGA analysis, carrageenan with 1% filler decomposes at 268 °C, while carrageenan with 2% filler decomposes at 278 °C. This can be attributed to the decrease in the exfoliated portion of graphites in the structure. That is, the graphene content of the composite decreases as the percentage increases. Alternatively, it is also conceivable that the fillers exceed the percolation threshold.

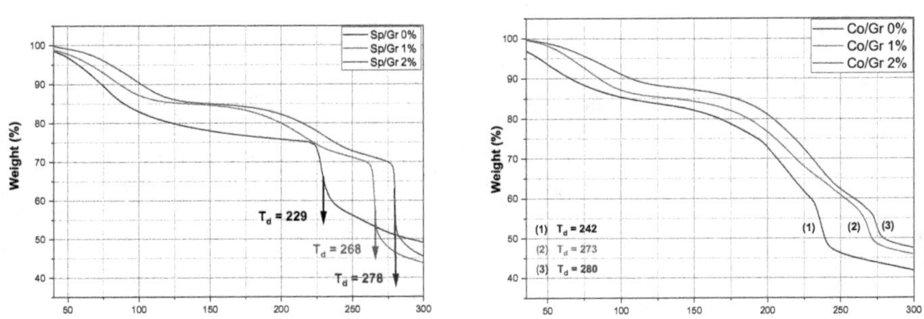

(a) (b)

Fig. 3. TGA curves of Sp/Gr(**a**) and Co/Gr(**b**)

According to the Fig. **3b**, the trends of *Co/Gr* (carrageenan from cottonii) are the same when compared with those of *Sp/Gr*. That is, at 1% loading the improvement is very significant, whereas at 2% the increase is relatively small. This can be explained in the same way as for *Sp/Gr*. And the small difference between the temperatures can also be explained by the small temperature difference of the matrices. Graphene show a higher thermal property than natural graphite due to the size effect. This study demonstrates the effectiveness of graphene nanosheets in enhancing the physicochemical properties of carrageenan matrices. The Fig. 2 and Fig. 3 shows the decomposition temperatures of the matrices(Carr-Co and Carr-Sp) and composites(*Co/Gr* and *Sp/Gr*). Carr-Co can be said to have a higher T_d (°C) than Carr-Sp, whether loaded or not.

3.2. *X-ray diffraction(XRD)*

Fig. 4 compares the diffractograms of graphite, the *Co/Gr* and the *Sp/Gr* films with different filler quantities. The spectra of Carr-Co and Carr-Sp show an amorphous structure, hence the absence of peaks in the *Co/Gr* 0% and *Sp/Gr* 0% films. The *Co/Gr* 1% film shows a basal reflection peak at $2\theta = 26.53°$ with a lower intensity than that of pure graphite, with a value shifted by 0.13°. The value of 2θ, can vary on the order of 0.2°.[24] The presence of this peak shows that graphite is incorporated throughout the composite. The low intensity explains why the area crossed by the graphite X-ray beams is greater than that of the graphite in the polymer. From

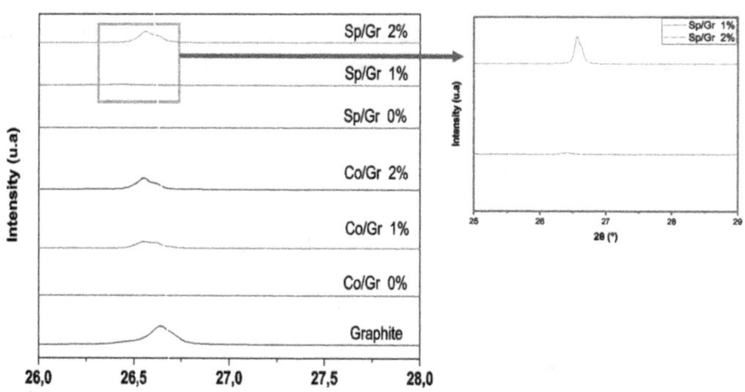

Fig. 4. XRD diagram of *Co/Gr* and *Sp/Gr*

the diffractograms, characteristic graphite diffraction peaks in carrageenan increase with increasing levels of incorporated graphite, but do not change angle. The first conclusion is that graphite does not exfoliate in the polymer. However, partial exfoliation cannot be ruled out. The XRD diagram of Sp/Gr 1% shows a peak at 2θ

= 26.40°, which characterises the graphite reflection and the hexagonal structure along the (200) orientation, as shown by the single peak, which also corresponds to an interreticular distance of $d_{200} = 0.34nm$ between the graphite planes.[1,25] The intensity of this peak is much lower than that of the Co/Gr 1% peak, even though they have the same charge ratio. This may reflect the reduction of graphite throughout the polymer, and may confirm the existence of graphite exfoliation giving rise to graphenes. Moreover, the appearance of the sample helps to prove this finding. The test has been carried out 5 times and always gives similar results.

3.3. *Fourier Transform Infrared(FT-IR) analysis*

The FT-IR spectra of Co/Gr and Sp/Gr are shown in Fig. 5. According to the figure, all peaks characteristic of carrageenan are present whether ι-carrageenan or κ-carrageenan. The peaks at 3336 and 1419 cm^{-1} respectively indicate the presence of the hydroxyl OH band ($CH_2 - OH$) and $C - H$ at alkane bending vibration CH_2.[26,27] The bands at 1223 cm^{-1} and 845 cm^{-1} correspond to $C{-}O{-}S$ stretching vibration and ester sulfate groups(symmetrical vibration $O = S = O$), respectively. In addition, the characteristic $C - O - C$ band of the anhydrous-galactose residue and the $C - O$ bridge stretching peak at are visible at 920 cm^{-1} and 1115 cm^{-1} respectively. The peak at 845 cm^{-1} confirms the stretching vibration of the $C - O - SO_3$ group.[28] Frequencies between 845 cm^{-1} is associated with the $C - S$ bond bending vibration of D-galactose-sulfate, which is related to the amount of sulfate in κ-carrageenan[29](Fig. 5**c**). The peak observed at 918 cm^{-1} corresponds to the $C - O - C$ of 3,6 anhydrogalactose. The peaks at 802 and 849 cm^{-1} are attributed to the $-O - SO_3$ stretching vibration of D-galactose-2-sulfate(DA2S) and D-galactose-4-sulfate(G4S).[30] The presence of these peaks confirms that we have ι-carrageenan, as Chitra et al have shown[31](Fig. 5**d**).

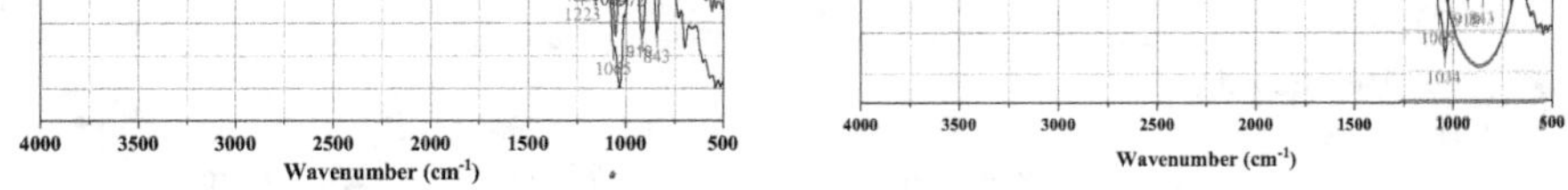

(c) (d)

Fig. 5. Spectre FIRT of Co/Gr(**c**) and Sp/Gr(**d**)

4. Conclusion

Carrageenan-based films differ in their physicochemical and thermal properties. Polymer production is based on the extraction of carrageenan from 2 types of red seaweed, cottonii and spinosum. The bionanocomposite films, on the other hand, were produced using a solution mixing and casting-evaporation technique.This approach yielded bio(nano)composite films with high-performance final properties, suitable for use in packaging and in the field of thermoplastic polymers. In the course of this work, a number of promising results have been achieved, but they remain preliminary and require further in-depth study. These results open up a number of prospects in the fields of nanomaterials based on red seaweed and graphite or graphene for enhancing the multifunctional properties of nanocomposites (biodegradable plastics).

Acknowledgments

We thank these institutes for their collaborations :

- Institut Halieutique et des Sciences Marines(IH.SM),University of Toliara
- Institut des Momécules et Materiaux du Mans, University of Mans(IMMM), France

ORCID

Rabearison Mizea MILLER `https://orcid.org/0009-0001-9175-0610`
Heriniaina.T. JOYEUX `https://orcid.org/0000-0001-7371-3560`
Maherizo Tiandrainy GEDICE `https://orcid.org/0009-0005-6719-0843`
Dimbinaina Olivier MAKA `https://orcid.org/0009-0000-1952-1939`

References

1. G. Wang, Yang *et al.*, *The Journal of Physical Chemistry C* **112**, 8192 (2008), DOI:https://doi.org/10.1021/jp710931h.
2. M. El Achaby, Nanocomposites graphène-polymère thermoplastique: Fabrication et étude des propriétés structurales, thermiques, rhéologiques et mécaniques, PhD thesis, Université Mohammed V-Agdal, Faculté des Sciences de Rabat; Faculté des ... (2012).
3. R. KAREK, Valorisation des matériaux composites d'origine naturelle, PhD thesis, Université Kasdi Merbah Ouargla (2022).
4. C. Scherba, Montreuil *et al.* (2018).
5. D. Chung, *Journal of materials science* **22**, 4190 (1987), DOI:https://doi.org/10.1007/BF01132008.
6. R. Croft, *Quarterly Reviews, Chemical Society* **14**, 1 (1960).
7. K. S. Novoselov, Geim *et al.*, *science* **306**, 666 (2004), DOI:https://doi.org/10.1126/science.1102896.
8. A. K. Geim and K. S. Novoselov, *Nature materials* **6**, 183 (2007), DOI:https://doi.org/10.1038/nmat1849.
9. Q. Wang, Ji *et al.*, *Molecules* **25**, 3306 (2020), DOI:https://doi.org/10.3390/molecules25143306.

10. A. A. Balandin, S. Ghosh, W. Bao, I. Calizo, D. Teweldebrhan, F. Miao and C. N. Lau, *Nano letters* **8**, 902 (2008), DOI:https://doi.org/10.1021/nl0731872.
11. G. Todinanahary, Behivoke *et al.*, *MHSA—PRU (Contrat n 166/C/PIC2/2016) Report. Toliara, Madagascar* (2016).
12. W. H. Organization *et al.* (2022).
13. M. Rinaudo, *Actualite Chimique* , 35 (2002).
14. M. Khotimchenko, V. Tiasto, A. Kalitnik, M. Begun, R. Khotimchenko, E. Leonteva, I. Bryukhovetskiy and Y. Khotimchenko, *Carbohydrate Polymers* **246**, 116568 (2020).
15. B. W. Souza, M. A. Cerqueira, A. I. Bourbon, A. C. Pinheiro, J. T. Martins, J. A. Teixeira, M. A. Coimbra and A. A. Vicente, *Food Hydrocolloids* **27**, 287 (2012).
16. H. Tsotetzo, Valorisation des polysaccharides marins: élaboration de nanocomposites et synthèse de graphène dopé, PhD thesis, Normandie Université (2017).
17. S. Toumi, Yahoum *et al.*, *Carbohydrate Polymers* **286**, 119310 (2022), DOI:https://doi.org/10.1016/j.carbpol.2022.119310.
18. P. Aurélie, Modification des propriétés physico-chimiques des carraghénanes par la purification et caractérisation de carraghénane-sulfatases, PhD thesis, Université Pierre et Marie Curie-UPMC (2012).
19. M. Abdorreza, Robal *et al.*, *LWT-Food Science and Technology* **46**, 135 (2012), DOI:https://doi.org/10.1016/j.lwt.2011.10.015.
20. J. T. Martins, Bourbon *et al.*, *Food and Bioprocess Technology* **6**, 2081 (2013), DOI:https://doi.org/10.1007/s11947-012-0851-4.
21. H. Jdidi, Fourati *et al.*, *Polymer* **228**, 123882 (2021), DOI:https://doi.org/10.1016/j.polymer.2021.123882.
22. X. Lu, Detrez *et al.*, *Composites Science and Technology* **213**, 108943 (2021), DOI:https://doi.org/10.1016/j.compscitech.2021.108943.
23. Y. Kanake, Développement de nanocomposites à base de polymères d'origine renouvelable-optimisation des propriétés barrière et de transport.— theses. fr, PhD thesis, Reims (2016).
24. A. Ibrahim, Klopocinska *et al.*, *Polymers* **13**, 2869 (2021), DOI:https://doi.org/10.3390/polym13172869.
25. A. Popova, *Coke and Chemistry* **60**, 361 (2017), DOI:https://doi.org/10.3103/S1068364X17090058.
26. N. A. A. Ghani, Othaman *et al.*, *Arabian journal of chemistry* **12**, 370 (2019), DOI:https://doi.org/10.1016/j.arabjc.2018.06.008.
27. A. Michel, M. Mestdagh and M. Axelos, *International Journal of Biological Macromolecules* **21**, 195 (1997), DOI:https://doi.org/10.1016/S0141-8130(97)00061-5.
28. M. D. Torres, Chenlo *et al.*, *Carbohydrate Polymers* **180**, 72 (2018), DOI:https://doi.org/10.1016/j.carbpol.2017.10.010.
29. C. Bellion (1979).
30. L. Pereira, Critchley *et al.*, *Journal of Applied Phycology* **21**, 599 (2009), DOI:https://doi.org/10.1007/s10811-009-9447-4.
31. R. Chitra, Sathya *et al.*, *Ionics* **25**, 2147 (2019), DOI:https://doi.org/10.1007/s11581-018-2687-z.

Polypropylene/Polystyrene and Clay nanocomposites : elaboration, morphological and thermal characterizations*

Heriniaina.T. JOYEUX[1]†, R.M. MILLER[1,2], B.T. OUDINOT[1], H. RISITE[1,2]

[1] *Laboratory of Physics and Environment, Physic Department University of Toliara PO Box 185, Toliara, Madagascar*

[2] *Test and Analysis of Materials, Institut Halieutique et des Sciences Marines(IH.SM), University of Toliara BP 141 Rue Dr. Rabesandratana, Mahavatsy II, 601 Toliara, Madagascar*

Received Day Month Year
Revised Day Month Year
Accepted Day Month Year
Published Day Month Year

The polypropylene (PP)/polystyrene (PS)/ montmorillonite (CA) clay blends were prepared by melt processing using PP-grafted silane (PP-g-VTEOS) as compatibilizer. At fixed content of clay (5 wt%), addition of PP-g-VTEOS was found to create, firstly, PS droplet well dispersed and distributed with reduced size proved by Scanning Electronic Microscopy (SEM) images and at the same time showed the improvement of the dispersion and distribution of clay. XRD spectra of nanocomposites revealed that intercalated/partial exfoliated structures were obtained. Thermogravimetric analysis showed that the degree of dispersion and distribution of clay within polymer matrix plays a vital role in property improvement.

Keywords: Immiscible PP/PS blends; Clay; Silane grafting; Nanocomposites; Thermal characterisation.

PACS numbers: 03.65.−w, 04.62.+v

1. Introduction

Poor compatibility between thermodynamically immiscible polymer blends leads to weak interfacial adhesion and unstable morphology. In order to enhance miscibility some interfacial stabilizers, known as compatibilizers are often used. Such a compatibilizer may be a homopolymer,[1] a graft, block or star copolymer and their choice depends on the nature of polymers and their characteristics.[2–4] The components of good polymer blends are selected in such a way that the primary

*Talk given at the 14th International Conference in High-Energy Physics - HEPMAD24 (21-26th October 2024, Antananarivo-Madagascar).

†Corresponding author: dadajoyeux@gmail.com

advantages of the first one would compensate the deficiencies of the second and vice versa considering their cost and the objectives.[5] The use of nanocharge was fund as one of ways to improve compatibility of polymer blends and leads to obtain high-performance polymeric materials.[6] Polypropylene (PP), polystyrene (PS) and their nanocomposites have found many applications in industry due to their low cost, easy processing and recyclability but PP/PS blends have poor compatibility due to the large differences in their polarities, chemical nature, and high interfacial tension. Generally, most immiscible polymer blends consist of droplets of a dispersed phase in a continuous phase, and these droplets may approach each other and finally coalesce.[7] Many researches were carried out focusing on compatibilization of PP/PS blends using different copolymers, reactive polymers, or nanoparticles as compatibilizers.[8–11] The most used compatibilizer polymer based is MA grafted polypropylene (PP-g-MA) derivative.[12,13] It consists of modifying polymer with a wide variety of polar groups leading to many outcomes such as enhancing compatibilizing effect or creating covalent bonds between two phases leading to better dispersion/distribution and favorable specific interactions. As nanoparticles, Ting Wu et al. used a nanocharge like montmorillonite clay to improve the interfacial adhesion and the strategy was found to reduce the PS droplet size and improve the compatibility of PP/PS.[6] To improve the interfacial adhesion of PP/PS blends the use of silane grafted PP (PP-g-Si) was not done yet.
Present work aims at preparing nanocomposites based on PP/PS/clay and compatibilized with vinyl triethoxysilane (VTEOS) by melt processing in twin-screw extruder. VTEOS has the advantage to play the role of compatibilizer. To the best of our knowledge, this is the first report on the use of silane to compatibilize the PP/PS blends. The aim of this study is therefore to investigate the effect of modified polymer on structural, morphological and thermal properties of the obtained nanocomposites and their relationship. The strategy of the preparation of the nanocomposites are presented in figure 1.

2. Materials and Methods

Organoclay, Cloiste 20 A, was obtained from Sothern Clay Products. The PP used throughout this work was a homopolymer, PP 5032E1, commercialized by Exxon Mobil chemical. Dicumyl peroxide (DCP, 98%), vinyl triethoxysilane (VTEOS, 97%) were from Sigma Aldrich. The dibutyltin dilaurate (DBTDL, 98%) was from Alfa Aesar. And the polystyrene (PS) used in this study was polystyrol 143 E purchased from D-BASF the chemical.

2.1. *Grafting procedure and nanocomposites preparation*

PP powder (50g) was tumble mixed with a solution of DCP (0,2 wt%) in C (5 wt%) for 20 min. The grafting reaction was carried out in Haake Polylab rheometer equipped with Rheomix mixing chamber and roller rotor. The mixture was reacted for 5min at 180 °Cat a rotation speed of 60 rpm, giving PP-g-VTEOS. The

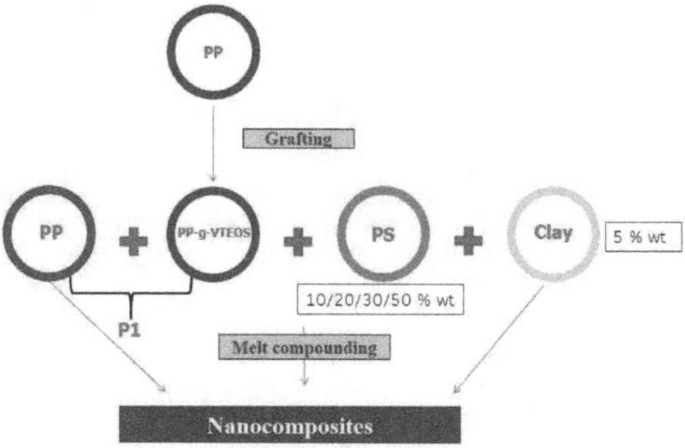

Fig. 1. Strategy of preparation of nanocomposites

characterization of functionalized PP is detailled in Ref.[13] Melt mixing of PP/PP-g-VTEOS/C20A and PP/PP-g-VTEOS/PS/C20A composites were carried out using contra-rotating twin screw micro extruder Minilab II. PP, PP-g-VTEOS, PS and nanoclay were tumble mixed and introduced in the hopper. 500 ppm of irganox-1010, 1000 ppm of Irgafos-168 and 600 ppm of calcium stearate were added after 5 min of mixing time for 2 min to stabilize PP-g-VTEOS for nanocomposites samples. The temperature, screw speed and residence time were fixed at 180 °C, 60 rpm and 5 min respectively. The composition of PP/PP-g-VTEOS is kept at 90 and 10 wt % for all blends and the clay concentrations was fixed at 5 wt %. The notation used the following for defining the blend composition is P1 for PP/PP-g-VTEOS, PS and C20A, expressed in weight fraction. The various compounds with their respective proportion are reported in Table 1.

2.2. *Structure and morphological characterization*

TThe XRD experiments were performed using X-ray scattering with a Bruker D8 AXS X-ray diffractometer. The angular range (2θ) was scanned from 1 to 10° at a step size of 0.05°, and the interlayer spacing (d_{200}) is calculated using the Bragg equation: $\lambda = 2d\sin\theta$, where λ is the wavelength of radiation (Cu-Kα radiation of 1.542 Å). SEM observations were made on ultrathin sections of nanocomposites prepared with a Leica EM UC7 ultramicrotome at room temperature, and then analyzed with a Philips CM200 TEM microscope operating at 200 kV, coupled with an EDS microprobe analyzer EDAX DX-4.

Table 1. Material designation and composition

Sample	Composition wt%
PP	100
PS	100
P1PS10	(90/10)/10
P1PS20	(90/10)/20
P1PS30	(90/10)/30
P1PS50	(90/10)/50
PP-PS10CA5	90/10/5
PP-PS20CA5	80/20/5
PP-PS30CA5	70/30/5
PP-PS50CA5	50/50/5
P1PS10CA5	[(90/10)/10]/5
P1PS20CA5	[(90/10)/20]/5
P1PS30CA5	[(90/10)/30]/5
P1PS50CA5	[(90/10)/50]/5

2.3. *Thermal characterization*

Thermogravimetric analysis, TGA, was performed on a Q500 TA instrument thermoanalyzer. The experiments were conducted at a temperature ramping of 10 °C/min from room temperature to 1000 °C, under air flow.

3. Results and discussion

The nanocomposites were elaborated by melt compounding method thanks to that method is closed to industrial production. The obtainend nanocomposites were subjected to morphological studies by X-Ray diffraction, scanning Electronic Microscopy and thermal stability by thermogravimetric analysis. The figure 2 presents the X-Ray diffraction patterns of the nanocomposites at different compositions of PS and compatibilizer. The virgin clay pattern showed a curve with a peak at 2θ of 3,6 equivalent to interplanar distance of 2,51 nm. However, all nanocomposites without and with compatibilizer at different rate of PS showed curves with peaks around 2θ of 2,4 equivalent to 3,54 nm. This shift is attributed to the intercalation of the polymeric chains into the clay cavity leading to the augmentation of the distance interfoliar. The results corresponds to the literature.[12,14] Remarkably, for nanocomposite P1PS10CA5, XRD spectra revealed a decrease in peak intensity which informed about a reduction of stacks intercalated structure leading to a higher degree of delamination. It means that the best degree of delamination is obtained at the low percentage of PS.

The morphological properties with SEM technics are presented in the figure 3 with selected samples. The SEM images of the lower content of PS without and with compatibilizer are compared at different magnifications. The images showed 3 significant points. First, it can be seen that PP and PS are immiscible. Secondly,

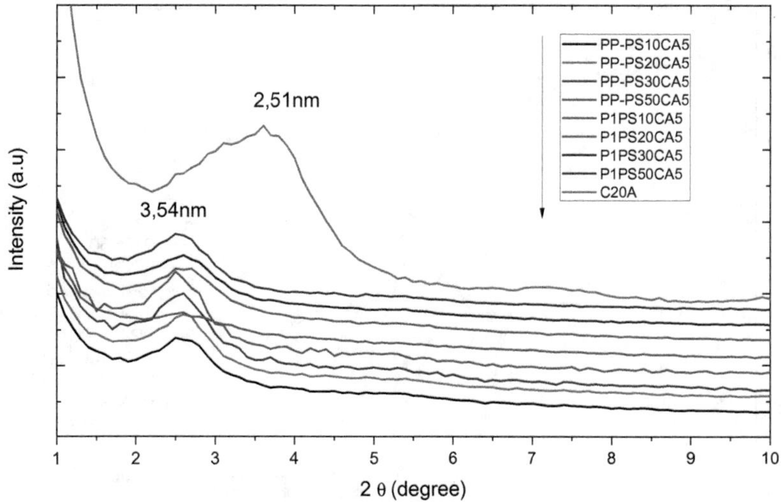

Fig. 2. XRD patterns for the clays and PP/PS nanocomposites at different contents of PP/PS

PS droplet well dispersed and distributed with reduced size were obtained with the addition of compatibilizer showing that its use affected the morphology of nanocomposites. Finally, the nanocomposites without compatibilizer showed disordered and random of PS phase and clay and it can be seen that the continuous phase doesn't contain clays. The presence of compatibilizer is important in morphological improvement properties. The last characterization studied in this work is the thermal stability of nanocomposites by TGA or thermogravimetric analysis. The figure 3 presents the mass loss of every composite in fonction of temperature. The studies showed that the nanocomposites with silane have better maximal decomposition temperature compared to those of without silane and the maximum is at the 50 % in PS about 430°C, an augmentation of about 100°C. The detailed data corresponding to the decomposition temperatures at 5% (T5%) and Tmax are presented in Table 2. The better dispersion and distribution of clay in the nanocomposites as told in morphological studies is responsible of these augmentation in terms of temperature. The clays acte like flame retardant which lead the decomposition temperature shifting to the higher.[15,16]

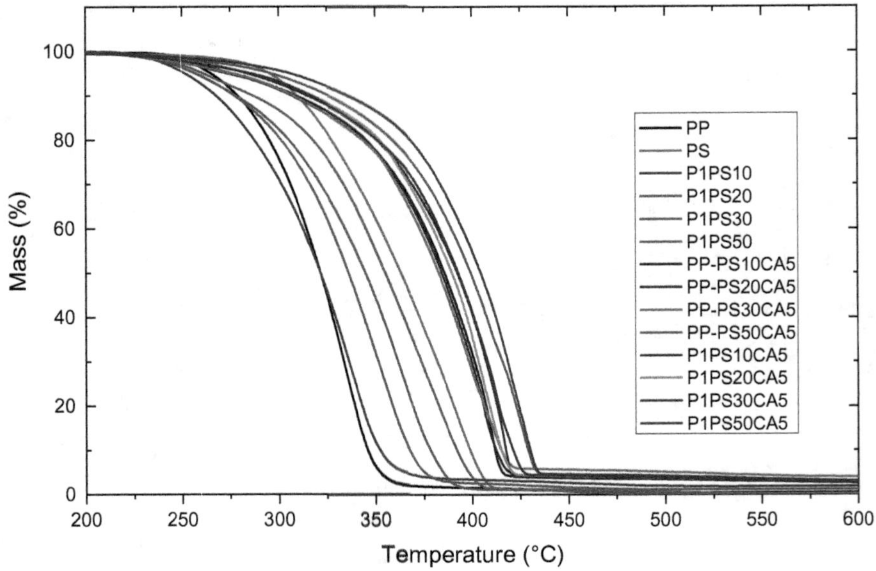

Fig. 3. TGA curves of pure PP, PS, and all nanocomposites at different contents of PP/PS.

Table 2. TGA data of pure PP and PS, PP/PP-g-VTEOS/PS and all nanocomposites

Sample	$T_{5\%}$(°C)	T_{max} (°C)
PP	265	334
PS	295	393
P1PS10	252	337
P1PS20	257	353
P1PS30	257	362
P1PS50	261	375
PP-PS10CA5	276	407
PP-PS20CA5	274	406
PP-PS30CA5	276	409
PP-PS50CA5	290	425
P1PS10CA5	284	414
P1PS20CA5	285	422
P1PS30CA5	290	425
P1PS50CA5	306	430

4. Conclusion

PP/PS clay based nanocomposites using PP-g-VTEOS were prepared by melt compounding. The use of PP-g-VTEOS promoted the improvement of propeties of

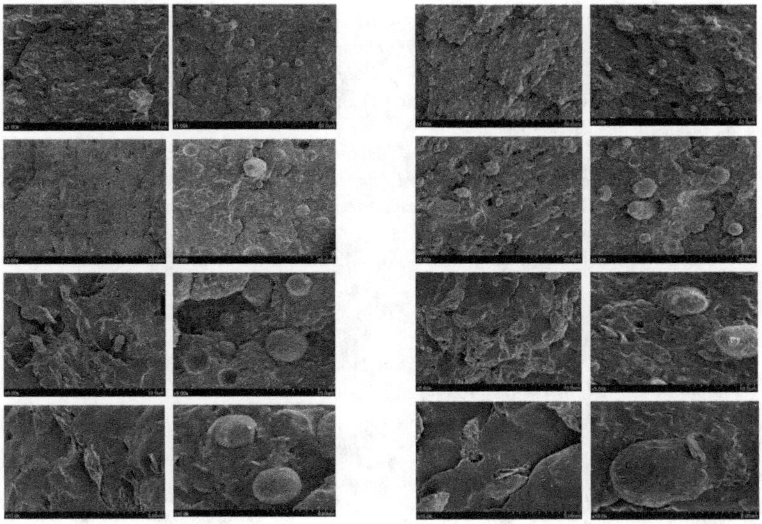

PPPS10CA5 vs P1PS10CA5 **PPPS50CA5 vs P1PS50CA5**

Fig. 4. Comparison SEM images of nanocomposites at different contents of PP/PS at different magnifications.(**PPPS10CA5 vs P1PS10CA5**) and *Sp/Gr*(**PPPS50CA5 vs P1PS50CA5**)

nanocomposites by the change in morphological structures. Despite the state of intercalation of clay nanocomposites, a noticeable increase was seen in the thermal properties, due to the better dispersion and distribution of clay and the creation of PS droplet with the reduced size. It can thus be expected that nanocomposites would have better properties only if the clay platelets were in the state of exfoliation.

Acknowledgments

This work was funded by MAScIR-Moroccan Foundation for Advanced Science, Innovation and Research. The authors are grateful for the financial support from EC (project REGPOT AL-NANOFUNC grant number REGPOTCT-2011-285895-Al-NANOFUNC-FP7).

ORCID

Heriniaina T. JOYEUX - `https://orcid.org/0000-0001-7371-3560`
MILLER Rabearison Mizea - `https://orcid.org/0009-0001-9175-0610`

References

1. J. Machado and C. Lee, *Polymer Engineering & Science* **34**, 59 (1994).
2. W. Barentsen, D. Heikens and P. Piet, *Polymer* **15**, 119 (1974).
3. B. Brahimi, A. Ait-Kadi, A. Ajji and R. Fayt, *Journal of Polymer Science Part B: Polymer Physics* **29**, 945 (1991).
4. D. Hlavata, Z. Horak, F. Lednický, J. Hromadkova, A. Pleska and Y. V. Zanevskii, *Journal of Polymer Science Part B: Polymer Physics* **39**, 931 (2001).
5. R. P. da Silva and C. P. Neto, *International Journal of Plastics Technology* **18**, 125 (2014).
6. T. Wu, D. Yuan, F. Qiu, R.-y. Chen, G.-z. Zhang and J.-p. Qu, *Polymer Testing* **63**, 73 (2017).
7. J.-M. Feng, X.-Q. Liu, R.-Y. Bao, W. Yang, B.-H. Xie and M.-B. Yang, *RSC advances* **5**, 74295 (2015).
8. Z. e. H. J. e. L. F. e. P. A. Hlavata, D et Horak, *Journal of Polymer Science Partie B : Physique des polym res* **37**, 1647.
9. P. H. P. Macaúbas and N. R. Demarquette, *Polymer* **42**, 2543 (2001).
10. R. Salehiyan, H. Y. Song, M. Kim, W. J. Choi and K. Hyun, *Macromolecules* **49**, 3148 (2016).
11. R. Salehiyan, Y. Yoo, W. J. Choi and K. Hyun, *Macromolecules* **47**, 4066 (2014).
12. G. Zhang, Y. Li and D. Yan, *Journal of Polymer Science Part B: Polymer Physics* **42**, 253 (2004).
13. H. Risite, H. A. Oualid and K. E. Mabrouk, Effects of vinyltriethoxysilane and maleic anhydride grafted polypropylenes on the morphological, thermal, rheological, and mechanical properties of polypropylene/clay nanocomposites, in *Proceedings*, (1) (2018), p. 6.
14. S. S. Nair, C. Ramesh and K. Tashiro, *Macromolecules* **39**, 2841 (2006).
15. J. Zhu, A. B. Morgan, F. J. Lamelas and C. A. Wilkie, *Chemistry of Materials* **13**, 3774 (2001).
16. B. Lepoittevin, M. Devalckenaere, N. Pantoustier, M. Alexandre, D. Kubies, C. Calberg, R. Jérôme and P. Dubois, *Polymer* **43**, 4017 (2002).

Part 7

Closing of the Conference

Acknowledgement and Future of the HEPMAD Conference by the Chairman of HEPMAD 24

Stephan Narison

Chairman of the HEPMAD Conference,

Directeur de Recherche Emerite CNRS au Laboratoire Univers et Particules de Montpellier,

CNRS & Université de Montpellier, Place Eugéne Bataillon, 34095-Montpellier Cedex 05, France

E-mail of the conference: qcdconf@gmail.com Url of the conference: https://www.lupm.umontpellier.fr/users/qcd/qcd20

Dear all,

I wish that everybody have had a good conference and spend a good week in Antananarivo during this meeting.

The sade news

HEPMAD 24 ends like the other previous HEPMAD conferences held in Antananarivo from 2001 to 2024.

The good news
is the announcement of the next meeting:

HEPMAD 25 is announced at the end of October 2025.

Indeed, we plan to maintain the organization of the HEPMAD Series of Conference as long as we can in order to keep activities in this field, to promote HEP and some other branches of physics.

In addition to the members of the committee, all participants are expected to be our ambassadors for stimulating participants to come to the next meeting.

I wish to thank the:

- Committee Members who continue to serve during several years,
- Spokespersons and Representants of different experiments.
- Hostesses for their efficient organization.
- Members of the iHEPMAD research institute
- Association Gasy Miara-Mandroso (AGMM) for the logistics.
- Participants because without them all these different organization efforts become useless.

I also wish that you have:

- Learned more about HEP and on some other branches of Physics and Chemistry,$\cdots$!

Finally, I wish that most of you will come back again (or send some of your colleagues) for presenting new results at the next HEPMAD 25 conference.

Have a good trip to your places !

Cordially,

Dr Stephan NARISON
QCD & HEP-MAD Conferences Chairman
Laboratoire Univers et Particules
CNRS / IN2P3 / UM2 Montpellier (FR)

Some Photos During HEPMAD 24

Some Photos of the Organizers

Official Opening of HEPMAD24

"Base Tulear" Debate

Some Scientific Sessions

Coffee - Breaks

Lunch - Breaks

Reserve Pereiras-Mandraka

Ambohimanga Palace

Folkloric Dances
Andry Mahitsy Group

Closing of Hepmad24

Participants

LIST OF PARTICIPANTS

Name	Institute	Email
DOMINIS PRESTER D.	Univ. Rijeka, Croatia (HV)	dijana.dominis@gmail.com
DOMINIS PRESTER P.	Univ. Rijeka, Croatia (HV)	pprester@uniri.hr
FANJIRINDRATOVO T.	Univ. Tulear (MG)	fanjiry@gmail.com
GEORGES R.	Univ. Tulear (MG)	georgesribiere1@gmail.com
JOYEUX H.T	Univ. Tulear (MG)	dadajoyeux@gmail.com
LIN S.-P.	INFN, Padova (IT)	lin@pd.infn.it
MANEFEMASY L.	Univ. Tulear (MG)	lmanefemasy@gmail.com
MILLER R.M	Univ. Tulear (MG)	rabearisonmiller@gmail.com
NARISON S.	CNRS-LUPM, Montpellier (FR)	snarison@yahoo.fr
PETERSEN T.	Univ. Copenhagen (DK)	petersen@nbi.dk
POPLAWSKI N.	Univ. New Haven (US)	NPoplawski@newhaven.edu
RABEMANANJARA A.	iHEPMAD, Antananarivo (MG)	achrisrab@gmail.com
RABETIARIVONY D.	iHEPMAD, Antananarivo (MG)	rd.bidds@gmail.com
RABOANARY R.	iHEPMAD, Antananarivo (MG)	r_raboanary@yahoo.fr
RAKOTOARINIRINA J.	iHEPMAD, Antananarivo (MG)	razoarinirina109@gmail.com
RAKOTONDRAINIBE M.	Collectif TANY (FR)	patrimoine.malagasy@gmail.com
RAKOTONIRINA H.	iHEPMAD, Antananarivo (MG)	hajarkt0@gmail.com
RAMAROSON J. F.	IST, Antananarivo (MG)	ramarosonflorence@gmail.com
RANAIVOSON R.T.	INSTN, Antananarivo (MG)	tokhiniaina@gmail.com
RANDRIAMARO Z.	CRAAD-OI, Antananarivo (MG)	randriamarozo@gmail.com
RANDRIANANDRASANA V.G.	Univ. Tulear (MG)	enamboare@yahoo.fr
RASOANAIVO A.	ENS, Antananarivo (MG)	phy.andry@yahoo.fr
RATSIFARITANA C.	ENS, Antananarivo (MG)	ratsifa@moov.mg
SETA D.	Univ. Tulear (MG)	seta.dahazizy@gmail.com
TOUSI M.S.	Univ. Teheran (IR)	marzieshekary@gmail.com
TSIMIHAFY R.J.	Univ. Tulear (MG)	trjacquelin@gmail.com
T.M. VAMI	UC-Santa Barbara (US)	Tamas.Almos.Vami@cern.ch

www.ingramcontent.com/pod-product-compliance
Lightning Source LLC
LaVergne TN
LVHW010335220926
848484LV00004BB/902